✔ KU-267-421

THE SCIENCE OF
STORYTELLING

GRIFFITH COLLEGE DUBLIN
SOUTH CIRCULAR ROAD DUBLIN 8.
Tel: 01 4150490 Fax: (01) 4549265
library@griffith.ie

GRIFFITH COLLEGE DUBLIN
3 9009 00070593 5

808.3
STO

Also by Will Storr

FICTION
The Hunger and the Howling of Killian Lone

NON FICTION
Will Storr Vs. The Supernatural
The Heretics: Adventures with the Enemies of Science
Selfie: How the West Became Self-Obsessed

COURSES
The Science of Storytelling
The Science of Storytelling for Business
How to Tell a Story

For more information about courses, please visit
www.scienceofstorytelling.co.uk

THE SCIENCE OF STORYTELLING

WILL STORR

GRIFFITH COLLEGE DUBLIN
SOUTH CIRCULAR ROAD DUBLIN 8.
Tel: 01 4150490 Fax: (01) 4549265
library@griffith.ie

WILLIAM
COLLINS

William Collins
An imprint of HarperCollins*Publishers*
1 London Bridge Street
London
SE1 9GF

www.WilliamCollinsBooks.com

First published in Great Britain by William Collins in 2019

5

Copyright © William Storr, 2019

William Storr asserts the moral right to be identified
as the author of this work in accordance with
the Copyright, Designs and Patents Act 1988

A catalogue record for this book is available from the British Library

ISBN 978-0-00-827693-5 (hardback)

Typeset in Freight Text by
Palimpsest Book Production Ltd, Falkirk, Stirlingshire

Printed and bound in Great Britain by
CPI Group (UK) Ltd, Croydon CRO 4YY

All rights reserved. No part of this publication may be reproduced,
stored in a retrieval system, or transmitted, in any form or by
any means, electronic, mechanical, photocopying, recording
or otherwise, without the prior permission of the publishers.

This book is sold subject to the condition that it shall not, by way of
trade or otherwise, be lent, re-sold, hired out or otherwise circulated
without the publisher's prior consent in any form of binding or cover
other than that in which it is published and without a similar condition
including this condition being imposed on the subsequent purchaser.

MIX
Paper from
responsible sources
FSC   **FSC® C007454**
www.fsc.org

This book is produced from independently certified FSC™ paper
to ensure responsible forest management.

For more information visit: www.harpercollins.co.uk/green

For my firstborn, Parker

'Ah, but a man's reach should exceed his grasp,
Or what's a heaven for?'

Robert Browning (1812–1889)

GRIFFITH COLLEGE DUBLIN
SOUTH CIRCULAR ROAD DUBLIN 8.
Tel: 01 4150490 Fax: (01) 4549265
library@griffith.ie

CONTENTS

CHAPTER TWO: THE FLAWED SELF

CHAPTER THREE: THE DRAMATIC QUESTION

INTRODUCTION

We know how this ends. You're going to die and so will everyone you love. And then there will be heat death. All the change in the universe will cease, the stars will die, and there'll be nothing left of anything but infinite, dead, freezing void. Human life, in all its noise and hubris, will be rendered meaningless for eternity.

But that's not how we live our lives. Humans might be in unique possession of the knowledge that our existence is essentially meaningless, but we carry on as if in ignorance of it. We beetle away happily, into our minutes, hours and days, with the fact of the void hovering over us. To look directly into it, and respond with an entirely rational descent into despair, is to be diagnosed with a mental-health condition, categorised as somehow faulty.

The cure for the horror is story. Our brains distract us from this terrible truth by filling our lives with hopeful goals and encouraging us to strive for them. What we want, and the ups and downs of our struggle to get it, is the story of us all. It gives our existence the illusion of meaning and turns our gaze from the dread. There's simply no way to understand the human world without stories. They fill our newspapers, our law courts, our sporting arenas, our government debating chambers, our school playgrounds, our computer games, the lyrics to our songs, our private thoughts and public conversations and our waking and sleeping dreams. Stories are everywhere. Stories are *us*.

It's story that makes us human. Recent research suggests language evolved principally to swap 'social information' back when we were living in Stone Age tribes. In other words, we'd gossip. We'd tell tales about the moral rights and wrongs of other people, punish the bad behaviour, reward the good, and thereby keep everyone cooperating and the tribe in check. Stories about people being heroic or villainous, and the emotions of joy and outrage they triggered, were crucial to human survival. We're wired to enjoy them.

Some researchers believe grandparents came to perform a vital role in such tribes: elders told different kinds of stories – about ancestor heroes, exciting quests and spirits and magic – that helped children to navigate their physical, spiritual and moral worlds. It's from these stories that complex human culture emerged. When we started farming and rearing livestock, and our tribes settled down and slowly merged into states, these grandparental campfire tales morphed into great religions that had the power to hold large numbers of humans together. Still, today, modern nations are principally defined by the stories we tell about our collective selves: our victories and defeats; our heroes and foes; our distinctive values and ways of being, all of which are encoded in the tales we tell and enjoy.

We experience our day-to-day lives in story mode. The brain creates a world for us to live in and populates it with allies and villains. It turns the chaos and bleakness of reality into a simple, hopeful tale, and at the centre it places its star – wonderful, precious *me* – who it sets on a series of goals that become the plots of our lives. Story is what brain does. It is a 'story processor', writes the psychologist Professor Jonathan Haidt, 'not a logic processor'. Story emerges from human minds as naturally as breath emerges from between human lips. You don't have to be a genius to master it. You're already doing it. Becoming better at telling stories is simply a matter of peering inwards, at the mind itself, and asking how it does it.

This book has an unusual genesis in that it's based on a storytelling course that is, in turn, based on research I've carried out for various books. My interest in the science of storytelling began about a decade ago when I was working on my second book, *The Heretics*, which was an investigation into the psychology of belief. I wanted to find out how intelligent people end up believing crazy things. The answer I found was that, if we're psychologically healthy, our brain makes us feel as if we're the moral heroes at the centre of the unfolding plots of our lives. Any 'facts' it comes across tend to be subordinate to that story. If these 'facts' flatter our heroic sense of ourselves, we're likely to credulously accept them, no matter how smart we think we are. If they don't, our minds will tend to find some crafty way of rejecting them. *The Heretics* was my introduction to the idea of the brain as a storyteller. It not only changed the way I saw myself, it changed the way I saw the world.

It also changed the way I thought about my writing. As I was researching *The Heretics*, I also happened to be working on my first novel. Having struggled with fiction for years I'd finally buckled and bought a selection of traditional 'how-to' guides. Reading through them, I noticed something odd. Some of the things the story theorists were saying about narrative were strikingly similar to what the psychologists and neuroscientists I'd been interviewing had been telling me about brain and mind. The storytellers and the scientists had started off in completely different places and had ended up discovering the same things.

As I continued my research, for subsequent books, I continued making these connections. I started to wonder if it might be possible to join the two fields up and thereby improve my own storytelling. That ultimately led to my starting a science-based course for writers which turned out to be unexpectedly successful. Being faced regularly with roomfuls of extremely smart authors, journalists and screen-writers pushed me to deepen my investigations. Soon, I realised I had about enough stuff to fill a short book.

My hope is that what follows will be of interest to anyone curious about the science of the human condition, even if they have little practical interest in storytelling. But it's also for the storytellers. The challenge any of us faces is that of grabbing and keeping the attention of other people's brains. I'm convinced we can all become better at what we do by finding out a bit about how they work.

This is an approach that stands in contrast to more traditional attempts at decoding story. These typically involve scholars comparing successful stories or traditional myths from around the world and working out what they have in common. From such techniques come predefined plots that put narrative events in a sequence, like a recipe. The most influential of these is undoubtedly Joseph Campbell's 'Monomyth', which, in its full form, has seventeen parts that track the phases of a hero's journey from their initial 'call to adventure' onwards.

Such plot structures have been hugely successful. They've drawn crowds of millions and dollars by the billions. They've led to an industrial revolution in yarn-spinning that's especially evident in cinema and long-form television. Some examples, such as the Campbell-inspired *Star Wars: A New Hope*, are wonderful. But too many more are Mars Bar stories, delicious and moreish but ultimately cold, corporate and cooked up by committee.

For me, the problem with the traditional approach is that it's led to a preoccupation with structure. It's easy to see why this has happened. Often the search has been for the One True Story – the ultimate, perfect plot structure by which every tale can be judged. And how are you going to describe *that* if not by dissecting it into its various movements?

I suspect it's this emphasis on structure that's responsible for the clinical feel from which many modern stories suffer. I believe the focus on plot should be shifted onto character. It's *people*, not events, that we're naturally interested in. It's the plight of specific, flawed and fascinating individuals that makes us cheer, weep and ram our heads into the sofa cushion. The surface events of the plot are crucial, of course, and structure ought to be present, functional and disciplined. But it's only there to support its cast.

While there are general structural principles, and a clutch of basic story shapes which are useful to understand, trying to dictate obligatory dos and don'ts that go beyond these extremely broad outlines is probably a mistake. A journey into the science of storytelling reveals that there are many things that attract and hold the attention of brains. Storytellers engage a number of neural processes that evolved for a variety of reasons and are waiting to be played like instruments in an orchestra: moral outrage, unexpected change, status play, specificity, curiosity, and so on. By understanding them, we can more easily create stories that are gripping, profound, emotional and original.

This, I hope, is an approach that will prove more creatively freeing. One benefit of understanding the science of storytelling is that it illuminates the 'whys' behind the 'rules' we're commonly given. Such knowledge should be empowering. Knowing *why* the rules are the rules means we know *how* to break them intelligently and successfully.

But none of this is to say we should disregard what story theorists such as Campbell have discovered. On the contrary. Many popular storytelling books contain brilliant insights about narrative and human nature that science has only recently caught up with. I quote a number of their authors in these pages. I'm not even arguing that we should ignore their valuable plot designs – they can easily be used to complement this book. It's really just a question of emphasis. I believe that compelling and unique plots are more likely to emerge from character than from a bullet-pointed list. And the best way to create characters that are rich and true and full of narrative surprise is to find out how characters operate in real life – and that means turning to science.

I've tried to write the storytelling book I wish I'd had, back when I was working on my novel. I've tried to balance *The Science of Storytelling* in such a way that it's of practical use without killing the creative spirit by issuing lists of 'You Musts'. I agree with the novelist and teacher of creative writing John Gardner, who argues that 'most supposed aesthetic absolutes prove relative under pressure'. If you're embarking on a storytelling project, I'd suggest you view what follows not as a series of obligations, but as weapons you can choose if and how to deploy. I've also outlined a practice that's proved successful in my classes over the years. The 'Sacred Flaw Approach' is a character-first process, an attempt to create a story that mimics the various ways a brain creates a life, and which therefore feels true and fresh, and comes pre-loaded with potential drama.

This book is divided into four chapters, each of which explores a different layer of storytelling. To begin, we'll examine how storytellers and brains create the vivid worlds they exist within. Next, we'll encounter the flawed protagonist at the centre of that world. Then we'll dive into that person's subconscious, revealing the hidden struggles and wills that make human life so strange and difficult, and the stories we tell about it so profound, compelling, unexpected and emotional. Finally, we'll be looking at the meaning and purpose of story and taking a fresh look at plots and endings.

What follows is an attempt to make sense of some of what generations of brilliant story theorists have discovered in the face of what equally brilliant women and men in the sciences have come to know. I am infinitely indebted to them all.

Will Storr

September, 2018

CHAPTER ONE:

CREATING A WORLD

———————————————

1.0

Where does a story begin? Well, where does anything begin? At the beginning, of course. Alright then: *Charles Foster Kane was born in Little Salem, Colorado, USA, in 1862. His mother was Mary Kane, his father was Thomas Kane. Mary Kane ran a boarding house . . ..*

It's not working. A birth may be the beginning of a life and, if the brain was a data processor, that's surely where our tale would start. But raw biographical data have little meaning to the storytelling brain. What it desires – what it insists upon, in exchange for the rare gift of its attention – is something else.

1.1

Many stories begin with a moment of unexpected change. And that's how they continue too. Whether it's a sixty-word tabloid piece about a TV star's tiara falling off or a 350,000-word epic such as *Anna Karenina*, every story you'll ever hear amounts to 'something changed'. Change is endlessly fascinating to brains. 'Almost all perception is based on the detection of change' says the neuroscientist Professor Sophie Scott. 'Our perceptual systems basically don't work unless there are changes to detect.' In a stable environment, the brain is relatively calm. But when it detects change, that event is immediately registered as a surge of neural activity.

It's from such neural activity that your experience of life emerges. Everything you've ever seen and thought; everyone you've loved and hated; every secret you've kept, every dream you've pursued, every sunset, every dawn, every pain, bliss, taste and longing – it's all a creative product of storms of information that loop and flow around your brain's distant territories. That 1.2-kg lump of pink computational jelly you keep between your ears might fit comfortably in two cupped hands but, taken on its own scale, it's vast beyond comprehension. You have 86 billion brain cells or 'neurons' and every one of them is as complex as a city. Signals flow between them at speeds of up to 120 metres per second. They travel along 150,000 to 180,000 kms of synaptic wiring, enough to wrap around the planet four times.

But what's all this neural power *for*? Evolutionary theory tells us our purpose is to survive and reproduce. These are complex aims, not least reproduction, which, for humans, means manipulating what potential mates think of us. Convincing a member of the opposite sex that we're a desirable mate is a challenge that requires a deep understanding of social concepts such as attraction, status, reputation and rituals of courting. Ultimately, then, we could say the mission of the brain is this: control. Brains have to perceive the physical environment and the people that surround it in order to *control* them. It's by learning how to control the world that they get what they want.

Control is why brains are on constant alert for the

unexpected. Unexpected change is a portal through which danger arrives to swipe at our throats. Paradoxically, however, change is also an opportunity. It's the crack in the universe through which the future arrives. Change is hope. Change is promise. It's our winding path to a more successful tomorrow. When unexpected change strikes we want to know, what does it mean? Is this change for the good or the bad? Unexpected change makes us curious, and curious is how we should feel in the opening movements of an effective story.

Now think of your face, not as a face, but as a machine that's been formed by millions of years of evolution for the detection of change. There's barely a space on it that isn't somehow dedicated to the job. You're walking down the street, thinking about nothing in particular, and there's unexpected change – there's a bang; someone calls your name. You stop. Your internal monologue ceases. Your powers of attention switch on. You turn that amazing change-detecting machine in its direction to answer the question, 'What's happening?'

This is what storytellers do. They create moments of unexpected change that seize the attention of their protagonists and, by extension, their readers and viewers. Those who've tried to unravel the secrets of story have long known about the significance of change. Aristotle argued that 'peripeteia', a dramatic turning point, is one of the most powerful moments in drama, whilst the story theorist and celebrated commissioner of screen drama John Yorke has written that 'the image every TV director in fact or fiction always looks for is the close-up of the human face as it registers change.'

These changeful moments are so important, they're often packed into a story's first sentences:

That Spot! He hasn't eaten his supper. Where can he be?
(*Eric Hill,* Where's Spot?)

Where's Papa going with that ax?
(*E. B. White,* Charlotte's Web)

When I wake up, the other side of the bed is cold.
(*Suzanne Collins,* The Hunger Games)

These openers create curiosity by describing specific moments of change. But they also hint darkly at troubling change to come. Could Spot be under a bus? Where *is* that man going with that axe? The threat of change is also a highly effective technique for arousing curiosity. The director Alfred Hitchcock, who was a master at alarming brains by threatening that unexpected change was looming, went as far as to say, 'There's no terror in the bang, only in the anticipation of it.'

But threatening change doesn't have to be as overt as a psycho's knife behind a shower curtain.

Mr and Mrs Dursley, of number four Privet Drive, were proud
to say that they were perfectly normal, thank you very much.
(*J. K. Rowling,* Harry Potter and the Philosopher's Stone)

Rowling's line is wonderfully pregnant with the threat of change. Experienced readers know *something* is about to pop the rather self-satisfied world of the Dursleys. This opener uses the same technique Jane Austen employs in *Emma*, which famously begins:

> *Emma Woodhouse, handsome, clever and rich, with a comfortable home and a happy disposition, seemed to unite some of the best blessings of existence; and had lived nearly twenty-one years in the world with very little to distress or vex her.*

As Austen's line suggests, using moments of change or the threat of change in opening sentences isn't some hack trick for children's authors. Here's the start of Hanif Kureishi's literary novel *Intimacy*:

> *It is the saddest night, for I am leaving and not coming back.*

Here's how Donna Tartt's *The Secret History* begins:

> *The snow in the mountains was melting and Bunny had been dead for several weeks before we came to understand the gravity of our situation.*

Here's Albert Camus starting *The Outsider*:

> *Mother died today. Or yesterday. I don't know.*

And here's Jonathan Franzen, opening his literary masterpiece *The Corrections* in precisely the same way that Eric Hill opened *Where's Spot*?

The madness of an autumn prairie cold front coming through. You could feel it: something terrible was going to happen.

Neither is it limited to modern story:

Rage! Sing, Goddess, [of] Achilles' rage, black and murderous, that cost the Greeks incalculable pain, pitched countless souls of heroes into Hades' dark, and left their bodies to rot as feasts for dogs and birds, as Zeus' will was done. Begin with the clash between Agamemnon, the Greek warlord, and godlike Achilles.   (*Homer,* The Iliad)

Or fiction:

A spectre is haunting Europe – the spectre of communism.
                      (*Karl Marx,* The Communist Manifesto)

And even when story a starts without much apparent change . . .

All happy families are alike; each unhappy family is unhappy in its own way.
                      (*Leo Tolstoy,* Anna Karenina *– first sentence.*)

. . . if it's going to earn the attention of masses of brains, you can bet change is on the way:

All was confusion in the Oblonskys' house. The wife had found out that the husband was having an affair with their former French governess and had announced to the husband that she could not live in the same house with him.

(*Leo Tolstoy,* Anna Karenina *– sentences two and three.*)

In life, most of the unexpected changes we react to will turn out to be of no importance: the bang was just a lorry door; it wasn't your name, it was a mother calling for her child. So you slip back into reverie and the world, once more, becomes a smear of motion and noise. But, every now and then, that change matters. It forces us to act. This is when story begins.

1.2

Unexpected change isn't the only way to arouse curiosity. As part of their mission to control the world, brains need to properly understand it. This makes humans insatiably inquisitive: between the ages of two and five, it's thought that we ask around 40,000 'explanatory' questions of our caregivers. Humans have an extraordinary thirst for knowing how things work and why. Storytellers excite these instincts by creating worlds but stopping short of telling readers everything about them.

The secrets of human curiosity have been explored by psychologists, perhaps most famously by Professor George Loewenstein. He writes of a test in which participants were confronted by a grid of squares on a computer screen. They were asked to click five of them. Some participants found that, with each click, another picture of an animal appeared. But a second group saw small component parts of a single animal. With each square they clicked, another part of a greater picture was revealed. This second group were much more likely to keep on clicking squares after the required five, and then keep going until enough of them had been turned that the mystery of the animal's identity had been solved. Brains, concluded the researchers, seem to become spontaneously curious when presented with an 'information set' they realise is incomplete. 'There is a natural inclination to resolve information gaps,' wrote Loewenstein, 'even for questions of no importance.'

Another study had participants being shown three photographs of parts of someone's body: hands, feet and torso. A second group saw two parts, a third saw one, while another group still saw none. Researchers found that the more photos of the person's body parts the participants saw, the greater was their desire to see a complete picture of the person. There is, concluded Loewenstein, a 'positive relationship between curiosity and knowledge'. The more context we learn about a mystery, the more anxious we become to solve it. As the stories reveal more of themselves, we increasingly want to know, Where *is* Spot? Who *is* 'Bunny' and *how* did he die and *how* is the narrator implicated in his death?

Curiosity is shaped like a lowercase n. It's at its weakest when people have no idea about the answer to a question and also when entirely convinced they do. The place of maximum curiosity – the zone in which storytellers play – is when people *think* they have *some* idea but aren't quite sure. Brain scans reveal that curiosity begins as a little kick in the brain's reward system: we crave to know the answer, or what happens next in the story, in the way we might crave drugs or sex or chocolate. This pleasantly unpleasant state, that causes us to squirm with tantalised discomfort at the delicious promise of an answer, is undeniably powerful. During one experiment, psychologists noted archly that their participants' 'compulsion to know the answer was so great that they were willing to pay for the information, even though curiosity could have been sated for free after the session.'

In his paper 'The Psychology of Curiosity', Loewenstein breaks down four ways of involuntarily inducing curiosity in humans: (1) the 'posing of a question or presentation of a puzzle'; (2) 'exposure to a sequence of events with an anticipated but unknown resolution'; (3) 'the violation of expectations that triggers a search for an explanation'; (4) knowledge of 'possession of information by someone else'.

Storytellers have long known these principles, having discovered them by practice and instinct. Information gaps create gnawing levels of curiosity in the readers of Agatha Christie and the viewers of *Prime Suspect*, stories in which they're (1) posed a puzzle; (2) exposed to a sequence of events with an anticipated but unknown resolution; (3) surprised by red herrings, and (4) tantalised by the fact that *someone* knows whodunnit, and how, but we don't. Without realising it, deep in the detail of his dry, academic paper, Loewenstein has written a perfect description of police-procedural drama.

It's not just detective stories that rely on information gaps. John Patrick Shanley's Pulitzer Prize-winning stage play *Doubt* toyed brilliantly with its audience's desire to know whether its protagonist, the avuncular and rebellious Catholic priest Father Flynn, was, in fact, a paedophile. The long-form journalist Malcolm Gladwell is a master at building curiosity about Loewensteinian 'questions of no importance' and manages the feat no more effectively than in his story 'The Ketchup Conundrum', in which he becomes a detective trying to solve the mystery of why it's so hard to make a sauce to rival Heinz.

Some of our most successful mass-market storytellers also rely on information gaps. J. J. Abrams is co-creator of the long-form television series *Lost*, which followed characters who mysteriously manage to survive an airline crash on a South Pacific island. There they discover mysterious polar bears; a mysterious band of ancient beings known as 'the Others'; a mysterious French woman; a mysterious 'smoke monster' and a mysterious metal door in the ground. Fifteen million viewers in the US alone were drawn to watch that first series, in which a world was created then filled until psychedelic with information gaps. Abrams has described his controlling theory of storytelling as consisting of the opening of 'mystery boxes'. Mystery, he's said, 'is the catalyst for imagination . . . what are stories but mystery boxes?'

1.3

In order to tell the story of your life, your brain needs to conjure up a world for you to live inside, with all its colours and movements and objects and sounds. Just as characters in fiction exist in a reality that's been actively created, so do we.

But that's not how it feels to be a living, conscious human. It *feels* as if we're looking out of our skulls, observing reality directly and without impediment. But this is not the case. The world we experience as 'out there' is actually a *reconstruction* of reality that is built *inside* our heads. It's an act of creation by the storytelling brain.

This is how it works. You walk into a room. Your brain predicts what the scene should look and sound and feel like, then it generates a hallucination based on these predictions. It's this hallucination that you experience as the world around you. It's this hallucination you exist at the centre of, every minute of every day. You'll never experience *actual* reality because you have no direct access to it. 'Consider that whole beautiful world around you, with all its colours and sounds and smells and textures,' writes the neuroscientist and fiction writer Professor David Eagleman. 'Your brain is not directly experiencing any of that. Instead, your brain is locked in a vault of silence and darkness inside your skull.'

This hallucinated reconstruction of reality is sometimes referred to as the brain's 'model' of the world. Of course, this model of what's actually out there needs to be somewhat accurate, otherwise we'd be walking into walls and ramming forks into our necks. For accuracy, we have our senses. Our senses seem incredibly powerful: our eyes are crystalline windows through which we observe the world in all its colour and detail; our ears are open tubes into which the noises of life freely tumble. But this is not the case. They actually deliver only limited and partial information to the brain.

Take the eye, our dominant sense organ. If you hold out

your arm and look at your thumbnail, that's all you can see in high definition and full colour at once. Colour ends 20 to 30 degrees outside that core and the rest of your sight is fuzzy. You have two lemon-sized blind spots and blink fifteen to twenty times a minute, which blinds you for fully 10 per cent of your waking life. You don't even see in three dimensions.

How is it, then, that we experience vision as being so perfect? Part of the answer lies in the brain's obsession with change. That large fuzzy area of your vision is sensitive to changes in pattern and texture as well as movement. As soon as it detects unexpected change, your eye sends its tiny high-definition core – which is a 1.5-millimetre depression in the centre of your retina – to inspect it. This movement – known as a 'saccade' – is the fastest in the human body. We make four to five saccades every second, over 250,000 in a single day. Modern filmmakers mimic saccadic behaviour when editing. Psychologists examining the so-called 'Hollywood style' find the camera makes 'match action cuts' to new salient details just as a saccade might, and is drawn to similar events, such as bodily movement.

The job of all the senses is to pick up clues from the outside world in various forms: lightwaves, changes in air pressure, chemical signals. That information is translated into millions of tiny electrical pulses. Your brain reads these electrical pulses, in effect, like a computer reads code. It uses that code to actively construct your reality, fooling you into believing this controlled hallucination is real. It then uses its senses as fact-checkers, rapidly tweaking what it's showing you whenever it detects something unexpected.

It's because of this process that we sometimes 'see' things that aren't actually there. Say it's dusk and you think you've seen a strange, stooping man with a top hat and a cane loitering by a gate, but you soon realise it's just a tree stump and a bramble. You say to your companion, 'I thought I saw a weird guy over there.' You *did* see that weird guy over there. Your brain thought he was there so it put him there. Then when you approached and new, more accurate, information was detected, it rapidly redrew the scene, and your hallucination was updated.

Similarly, we often *don't* see things that *are* actually there. A series of iconic experiments had participants watch a video of people throwing a ball around. They had to count the number of times the ball was passed. Half didn't spot a man in a gorilla suit walk directly into the middle of the screen, bang his chest three times, and leave after fully nine seconds. Other tests have confirmed we can also be 'blind' to auditory information (the sound of someone saying 'I am a gorilla' for nineteen seconds) as well as touch and smell information. There's a surprising limit to how much our brains can actually process. Pass that limit and the object is simply edited out. It's not included in our hallucinated reality. It literally becomes invisible to us. These findings have dire potential consequences. In a test of a simulated vehicle stop, 58 per cent of police trainees and 33 per cent of experienced officers 'failed to notice a gun positioned in full view on the passenger dashboard'.

Things naturally become worse when our fact-checking senses become damaged. When people's eyesight develops sudden flaws, their hallucinatory model of reality can begin to flicker and fail. They sometimes see clowns, circus animals and cartoon characters in the areas that have gone dark. Religious people have apparent visitations. These individuals are not 'mad' and neither are they rare. The condition affects millions. Dr Todd Feinberg writes of a patient, Lizzy, who suffered strokes in her occipital lobes. As can happen in such cases, her brain didn't immediately process the fact she'd gone 'suddenly and totally' blind, so it continued projecting its hallucinated model of the world. Visiting her hospital bed, Feinberg enquired if she was having trouble with her vision in any way. 'No,' she said. When he asked her to take a look around and tell him what she saw, she moved her head accordingly.

'It's good to see friends and family, you know,' she said. 'It makes me feel like I'm in good hands.'

But there was nobody else there.

'Tell me their names,' said Feinberg.

'I don't know everybody. They're my brother's friends.'

'Look at me. What am I wearing?'

'A casual outfit. You know, a jacket and pants. Mostly navy blue and maroon.'

Feinberg was in his hospital whites. Lizzy continued their chat smiling and acting 'as if she had not a care in the world'.

These relatively recent findings by neuroscientists demand a spooky question. If our senses are so limited, how do we know what's actually happening outside the dark vault of our skulls? Disturbingly, we don't know for sure. Like an old television that can only pick up black and white, our biological technology simply can't process most of what's actually going on in the great oceans of electromagnetic radiation that surround us. Human eyes are able to read less than one ten-trillionth of the light spectrum. 'Evolution shaped us with perceptions that allow us to survive,' the cognitive scientist Professor Donald Hoffman has said. 'But part of that involves hiding from us the stuff we don't need to know. And that's pretty much all of reality, whatever reality might be.'

We do know that actual reality is radically different than the model of it that we experience in our heads. For instance, there's no sound out there. If a tree falls in a forest and there's no one around to hear it, it creates changes in air pressure and vibrations in the ground. The crash is an effect that happens in the brain. When you stub your toe and feel pain throbbing out of it, that, too, is an illusion. That pain is not in your toe, but in your brain.

There's no colour out there either. Atoms are colourless. All the colours we do 'see' are a blend of three cones that sit in the eye: red, green and blue. This makes us *Homo sapiens* relatively impoverished members of the animal kingdom: some birds have six cones; mantis shrimp have *sixteen*; bees' eyes are able to see the electromagnetic structure of the sky. The colourful worlds they experience beggar human imagination. Even the colours we do 'see' are mediated by culture. Russians are raised to see two types of blue and, as a result, see eight-striped rainbows. Colour is a lie. It's set-dressing, worked up by the brain. One theory has it that we began painting colours onto objects millions of years ago in order to identify ripe fruit. Colour helps us interact with the external world and thereby better control it.

The only thing we'll ever really know are those electrical pulses that are sent up by our senses. Our storytelling brain uses those pulses to create the colourful set in which to play out our lives. It populates that set with a cast of actors with goals and personalities, and finds plots for us to follow. Even sleep is no barrier to the brain's story-making processes. Dreams feel real because they're made of the same hallucinated neural models we live inside when awake. The sights are the same, the smells are the same, objects feel the same to the touch. Craziness happens partly because the fact-checking senses are offline, and partly because the brain has to make sense of chaotic bursts of neural activity that are the result of our state of temporary paralysis. It explains this confusion as it explains everything: by roughing together a model of the world and magicking it into a cause-and-effect story.

One common dream has us falling off a building or tumbling down steps, a brain story that's typically triggered to explain a 'myoclonic jerk', a sudden, jarring contraction of the muscles. Indeed, just like the stories we tell each other for fun, dream narratives often centre on dramatic, unexpected change. Researchers find the majority of dreams feature at least one event of threatening and unexpected change, with most of us experiencing up to five such events every night. Wherever studies have been done, from East to West, from city to tribe, dream plots reflect this. 'The most common is being chased or attacked,' writes story psychologist Professor Jonathan Gottschall. 'Other universal themes include falling from a great height, drowning, being lost or trapped, being naked in public, getting injured, getting sick or dying, and being caught in a natural or manmade disaster.'

So now we've discovered how reading works. Brains take information from the outside world – in whatever form they can – and turn it into models. When our eyes scan over letters in a book, the information they contain is converted into electrical pulses. The brain reads these electrical pulses and builds a model of whatever information those letters provided. So if the words on the page describe a barn door hanging on one hinge, the reader's brain will model a barn door hanging on one hinge. They'll 'see' it in their heads. Likewise, if the words describe a ten-foot wizard with his knees on back to front, the brain will model a ten-foot wizard with his knees on back to front. Our brain rebuilds the model world that was originally imagined by the author of the story. This is the reality of Leo Tolstoy's brilliant assertion that 'a real work of art destroys, in the consciousness of the receiver, the separation between himself and the artist.'

A clever scientific study examining this process seems to have caught people in the act of 'watching' the models of stories that their brains were busily building. Participants wore glasses that tracked their saccades. When they heard stories in which lots of events happened above the line of the horizon, their eyes kept making micro-movements upwards, as if they were actively scanning the models their brains were generating of its scenes. When they heard 'downward' stories, that's where their eyes went too.

The revelation that we experience the stories we read by

building hallucinated models of them in our heads makes sense of many of the rules of grammar we were taught at school. For the neuroscientist Professor Benjamin Bergen, grammar acts like a film director, telling the brain what to model and when. He writes that grammar 'appears to modulate what part of an evoked simulation someone is invited to focus on, the grain of detail with which the simulation is performed, or what perspective to perform that simulation from'.

According to Bergen, we start modelling words as soon as we start reading them. We don't wait until we get to the end of the sentence. This means the order in which writers place their words matters. This is perhaps why transitive construction – *Jane gave a Kitten to her Dad* – is more effective than the ditransitive – *Jane gave her Dad a kitten*. Picturing Jane, then the Kitten, then her Dad mimics the real-world action that we, as readers, should be modelling. It means we're mentally experiencing the scene in the correct sequence. Because writers are, in effect, generating neural movies in the minds of their readers, they should privilege word order that's filmic, imagining how their reader's neural camera will alight upon each component of a sentence.

For the same reason, active sentence construction – *Jane kissed her Dad* – is more effective than passive – *Dad was kissed by Jane*. Witnessing this in real life, Jane's initial movement would draw our attention and then we'd watch the kiss play out. We wouldn't be dumbly staring at Dad, waiting for something to happen. Active grammar means readers model the scene on the page in the same way that they'd model it if it happened in front of them. It makes for easier and more immersive reading.

A further powerful tool for the model-creating storyteller is the use of specific detail. If writers want their readers to properly model their story-worlds they should take the trouble to describe them as precisely as possible. Precise and specific description makes for precise and specific models. One study concluded that, to make vivid scenes, three specific qualities of an object should be described, with the researcher's examples including 'a dark blue carpet' and 'an orange striped pencil.'

The findings Bergen describes also suggest the reason writers are continually encouraged to 'show not tell'. As C. S. Lewis implored a young writer in 1956, 'instead of telling us a thing was "terrible", describe it so that we'll be terrified. Don't say it was "delightful"; make us say "delightful" when we've read the description.' The abstract information contained in adjectives such as 'terrible' and 'delightful' is thin gruel for the model-building brain. In order to experience a character's terror or delight or rage or panic or sorrow, it has to make a model of it. By building its model of the scene, in all its vivid and specific detail, it experiences what's happening on the page almost as if it's actually happening. Only that way will the scene truly rouse our emotions.

Mary Shelley may have been a teenager writing more than 170 years before the discovery of our model-making processes, but when she introduces us to Frankenstein's monster she displays an impressive instinct for its ramifications: filmic word order; specificity and show-not-tell.

It was already one in the morning; the rain pattered dismally against the panes, and my candle was nearly burned out, when, by the glimmer of the half-extinguished light, I saw the dull yellow eye of the creature open; it breathed hard, and a convulsive motion agitated its limbs. How can I describe my emotions at this catastrophe, or how delineate the wretch whom with such infinite care and pains I had endeavoured to form? His limbs were in proportion, and I had selected his features as beautiful. Beautiful! Great god! His yellow skin scarcely covered the work of muscles and arteries beneath; his hair was of a lustrous black, and flowing; his teeth was of a pearly whiteness; but these luxuriances only formed a more horrid contrast with his watery eyes, that seemed almost of the same colour as the dun-white sockets in which they were set, his shrivelled complexion and straight black lips.

Immersive model worlds can also be summoned by the evocation of the senses. Touches, tastes, scents and sounds can be recreated in the brains of readers as the neural networks associated with these sensations become activated when they see the right words. All it takes is deployment of specific detail, with the sensory information ('a cabbagey') paired to visual information ('brown sock'). This simple technique is used to magical effect in Patrick Süskind's novel *Perfume*. It tells of an orphan with an awesome sense of smell who's born in a malodorous fish market. He takes us into his world of eighteenth-century Paris by conjuring a kingdom of scent:

the streets stank of manure, the courtyards of urine, the stairwells stank of mouldering wood and rat droppings, the kitchens of spoiled cabbage and mutton fat; the unaired parlours stank of stale dust, the bedrooms of greasy sheets, damp featherbeds and the pungently sweet aroma of chamber-pots. The stench of sulphur rose from the chimneys, the stench of caustic lyes from the tanneries, and from the slaughterhouses came the stench of congealed blood. People stank of sweat and unwashed clothes; from their mouths came the stench of rotting teeth, from their bellies that of onions, and from their bodies, if they were no longer very young, came the stench of rancid cheese and sour milk and tumorous disease . . . [the heat of day squeezed] its putrefying vapour, a blend of rotting melon and the fetid odour of burned animal horn, out into the nearby alleys.

1.4

The brain's propensity for automatic model-making is exploited with superb effect by tellers of fantasy and science-fiction stories. Simply naming a planet, ancient war or obscure technical detail seems to trigger the neural process of building it, as if it actually exists. One of the first books I fell in love with as a boy was J.R.R. Tolkien's *The Hobbit*. My best friend Oliver and I obsessed over the maps it contained – 'Mount Gundabad'; 'Desolation of Smaug'; 'West lies Mirkwood the Great – there are spiders.' When his father made photocopies of them for us, these maps became the focus of a summer of blissful play. The places Tolkien sketched out, on those maps, felt as real to us as the sweet shop in Silverdale Road.

In *Star Wars*, when Han Solo boasts that his ship the *Millennium Falcon* 'made the Kessel Run in less than twelve parsecs' we have the strange experience of knowing it's an actor doing gibberish whilst simultaneously somehow *feeling* as if it's real. The line works because of its absolute specificity and its adherence to what sounds like truth (the 'Kessel Run' really could be a race while 'parsecs' are a genuine measurement of distance, equivalent to 3.26 light years). As ridiculous as some of this language actually is, rather than taking us out of the storyteller's fictional hallucination, it manages to give it even more density.

By merest suggestion, the Kessel Run becomes real. We can imagine the dusty planet on which the race begins, hear the whine and blast of the engines, smell the alien piss around the back of the mechanics' wind-flapping encampments. This is just what happens in *Bladerunner*'s most famous scene, in which the replicant Roy Batty, on the edge of death, tells Rick Deckard, 'I've seen things you people wouldn't believe. Attack ships on fire off the shoulder of Orion. I watched C-beams glitter in the dark near the Tannhäuser Gate.'

Those C-beams! That gate! Their wonder lies in the fact that they're merely suggested. Like monsters in the most frightening horror stories, they feel all the more real for being the creations, not of the writer, but of our own incessant model-making imaginations.

1.5

The hallucinated world our brain creates for us is specialised. It's honed towards our particular survival needs. Like all animals, our species can only detect the narrow band of reality that's necessary for us to get by. Dogs live principally in a world of smell, moles in touch and knife-fish in a realm of electricity. The human world is predominantly that of people. Our hyper-social brains are designed to control an environment of other selves.

Humans have an extraordinary gift for reading and understanding the minds of other people. In order to control our environment of humans, we have to be able to predict what they're going to do. The importance and complexity of human behaviour means we have an insatiable curiosity about it. Storytellers exploit both these mechanisms and this curiosity; the stories they tell are a deep investigation into the ever-fascinating whys of what people do.

We've been a social species, whose survival has depended upon human cooperation, for hundreds of thousands of years. But over the last 1,000 generations it's been argued that these social instincts have been rapidly honed and strengthened. This 'sharp acceleration' of selection for social traits, writes developmental psychologist Professor Bruce Hood, has left us with brains that are 'exquisitely engineered to interact with other brains'.

For earlier humans that roamed hostile environments, aggression and physicality had been critical. But the more cooperative we became, the less useful these traits proved.

When we started living in settled communities, they grew especially troublesome. There, it would've been the people who were better at getting along with others, rather than the physically dominant, who'd have been more successful.

This success in the community would've meant greater reproductive success, which would've gradually led to the emergence of a new strain of human. These humans had thinner and weaker bones than their ancestors and greatly reduced muscle mass, their physical strength as much as halving. They also had the kind of brain chemistry and hormones that predisposed them to behaviour specialised for settled communal living. They'd have been less interpersonally aggressive, but more adept at the kind of psychological manipulation necessary for negotiating, trading and diplomacy. They'd become expert at controlling their environment of other human minds.

You might compare it to the difference between a wolf and a dog. A wolf survives by cooperating as well as fighting for dominance and killing prey. A dog does so by manipulating its human owner such that they'd do *anything* for them. The power my beloved labradoodle Parker has over my own brain is frankly embarrassing. (I've dedicated this bloody *book* to her.) In fact, this might be more than a mere analogy. Researchers such as Hood argue that modern humans, just like dogs, have gone through a process of domestication. Support for the idea comes partly from the fact that, over the last 20,000 years, our brains have shrunk by between ten and fifteen per cent, the same reduction that's been observed in all the thirty or so other animals that humans have domesticated. Just as with those creatures, our domestication means we're tamer than our ancestors, better at reading social signals and more dependent on others. But, writes Hood, 'no other animal has taken domestication to the extent that we have.' Our brains may have initially evolved to 'cope with a potentially threatening world of predators, limited food and adverse weather, but we now rely on it to navigate an equally unpredictable social landscape.'

Unpredictable humans. This is the stuff of story.

For modern humans, controlling the world means controlling other people, and that means understanding them. We're wired to be fascinated by others and get valuable information from their faces. This fascination begins almost immediately. Whereas ape and monkey parents spend almost no time looking at their babies' faces, we're helplessly drawn to them. Newborns are attracted to human faces more than to any other object and, one hour from birth, begin imitating them. By two, they've learned to control their social worlds by smiling. By the time they're adults, they've become so adept at reading people that they're making calculations about status and character automatically, in one tenth of a second. The evolution of our strange, extremely other-obsessed brains has brought with it weird side-effects. Human obsession with faces is so fierce we see them almost anywhere: in fire; in clouds; down spooky corridors; in toast.

We sense minds everywhere too. Just as the brain models the outside world it also builds models of minds. This skill, which is an essential weapon in our social armoury, is known as 'theory of mind'. It enables us to imagine what others are thinking, feeling and plotting, even when they're not present. We can experience the world from another's perspective. For the psychologist Professor Nicholas Epley this capacity, which is obviously essential for storytelling, gave us incredible power. 'Our species has conquered the Earth because of our ability to understand the minds of others,' he writes, 'not because of our opposable thumbs or handiness with tools.' We develop this skill at around the age of four. It's then that we become story-ready; equipped to understand the logic of narrative.

The human ability to populate our minds with imagined other minds is the start of religion. Shamans in hunter-gatherer tribes would enter trance states and interact with spirits, and use these interactions as attempts to control the world. Religions were also typically animistic: our storytelling brains would project human-like minds into trees, rocks, mountains and animals, imagining they were possessed by gods who were responsible for changeful events, and required controlling with ritual and sacrifice.

Childhood stories reflect our natural tendency for such hyperactive mind-detecting. In fairytales, human-like minds are everywhere: mirrors talk, pigs eat breakfast, frogs turn into princes. Youngsters naturally treat their dolls and teddies as if they're inhabited by selves. I remember feeling terrible guilt for preferring my pink bear, handmade by my Grandmother, to my shop-bought brown bear. I knew they both knew how I felt, and that left me distracted and sad.

We never really grow out of our inherent animism. Which one of us hasn't kicked a door that's slammed on our fingers believing, in that disorientating flash of pain, that it attacked us out of spite? Who among us hasn't told a self-assembly wardrobe to fuck off? Whose storytelling brain doesn't commit its own literary-style pathetic fallacy, allowing the sun to make them optimistic about the coming day or the brooding clouds pessimistic? Studies indicate that those who anthropomorphise a human personality onto their cars show less interest in trading them. Bankers project human moods onto the movements of the markets and place their trades accordingly.

When we're reading, hearing or watching a story we deploy our theory-of-mind skills by automatically making hallucinatory models of the minds of its characters. Some authors model the minds of their own characters with such force that they hear them talk. Charles Dickens, William Blake and Joseph Conrad all spoke of such extraordinary experiences. The novelist and psychologist Professor Charles Fernyhough has led research in which 19 per cent of ordinary readers reported hearing the voices of fictional characters even after they'd put their books down. Some reported a kind of literary possession, with the character influencing the tone and nature of their thoughts.

But much as humans excel at such feats of theory of mind, we also tend to dramatically overestimate our abilities. Although there's an admitted absurdity in claiming to be able to quantify human behaviour with such absolute numerical precision, some research suggests strangers read another's thoughts and feelings with an accuracy of just 20 per cent. Friends and lovers? A mere 35 per cent. Our errors about what others are thinking are a major cause of human drama. As we move through life, wrongly predicting what people are thinking and how they'll react when we try to control them, we haplessly trigger feuds and fights and misunderstandings that fire devastating spirals of unexpected change into our social worlds.

Comedy, whether by William Shakespeare or John Cleese and Connie Booth, is often built on such mistakes. But whatever the mode of storytelling, well-imagined characters always have theories about the minds of other characters and – because this is drama – those theories will often be wrong. This wrongness will lead to unexpected consequences and yet more drama. The influential post-war director Alexander Mackendrick writes, 'I start by asking: What does A think B is thinking about A? It sounds complicated (and it is) but this is the very essence of giving some density to a character and, in turn, a scene.'

The author Richard Yates uses a theory-of-mind mistake to create a pivotal moment of drama in his classic *Revolutionary Road*. The novel charts the dissolving marriage of Frank and April Wheeler. When they were young, and newly in love, Frank and April dreamed of bohemian lives in Paris. But, when we meet them, middle-aged reality has struck. Frank and April have two children, with a third on the way, and have moved into a cookie-cutter suburb. Frank's secured a job at his father's old company and has found himself rather settling into a life of boozy lunches and housewife-at-home ease. But April isn't happy. She still dreams of Paris. They argue, bitterly. Sex is withheld. Frank sleeps with a girl at work. And then he makes his theory-of-mind mistake.

In order to break the impasse with his wife, Frank decides to confess his infidelity. His theory of April's mind appears to be that she'll be thrown into a state of catharsis that will jolt her back into reality. There'll be tears to mop up, sure, but those tears will just remind the ol' gal why she loves him.

This is not what happens. When he confesses, April asks, *Why?* Not why he slept with the girl, but why is he bothering to tell her? She doesn't care about his fling. This isn't what Frank was expecting at all. He wants her to care! 'I know you do,' April tells him. 'And I suppose I would, if I loved you; but you see I don't. I don't love you and I never really have and I never really figured it out until this week.'

1.6

As the eye darts about, building up its story world for you to live inside, the brain's choosy about where it tells it to look. We're attracted to change, of course, but also to other salient details. Scientists used to believe attention was drawn simply to objects that stood out, but recent research suggests we're more likely to attend to that which we find meaningful. Unfortunately, it's not yet known precisely what 'meaningful' means, in this context, but tests that tracked saccades found, for example, that an untidy shelf attracted more attention than a sun-splashed wall. For me, that untidy shelf hints of human change; of a life in detail; of trouble insinuating itself in a place designed for order. It's no surprise test-brains were drawn to it. It's story-stuff, whilst the sun is just a shrug.

Storytellers also choose carefully what meaningful details

to show and when. In *Revolutionary Road*, just after Frank makes his changeful theory-of-mind mistake that throws his life in a new and unexpected direction, the author draws our attention to one brilliant detail. It's an urgent voice on the radio: 'And listen to this. Now, during the Fall Clearance, you'll find Robert Hall's *entire stock* of men's walk shorts and sport jeans drastically reduced!'

Both believable and crushing, it serves to intensify our feelings, at exactly the right moment, of the suffocating and dreary housewifey corner that April has found herself backed into. Its timing also implicitly defines and condemns what Frank has become. He used to think he was bohemian – a thinker! – and now he's just Bargain Shorts Man. This is an advert for him.

The director Stephen Spielberg is famous for his use of salient detail to create drama. In *Jurassic Park*, during a scene that builds to our first sighting of *Tyrannosaurus rex*, we see two cups of water on a car dashboard, deep rumbles from the ground sending rings over their liquid surface. We cut between the faces of the passengers, each slowly registering change. Then we see the rear-view mirror vibrating with the stomping of the beast. Extra details like this add even more tension by mimicking the way brains process peak moments of stress. When we realise our car is about to crash, say, the brain needs to temporarily increase its ability to control the world. Its processing power surges and we become aware of more features in our environment, which has the effect of making time seem to slow down. In exactly this way, storytellers stretch time, and thereby build suspense, by packing in extra saccadic moments and detail.

1.7

There's a park bench, in my hometown, that I don't like to walk past because it's haunted by a breakup with my first love. I see ghosts on that bench that are invisible to anyone else except, perhaps, her. And I feel them too. Just as human worlds are haunted with minds and faces, they're haunted with memories. We think of the act of 'seeing' as the simple detection of colour, movement and shape. But we see with our pasts.

That hallucinatory neural model of the world we live inside is made up of smaller, individual models – we have neural models of park benches, dinosaurs, ISIS, ice cream, models of *everything* – and each of those is packed with associations from our own personal histories. We see both the thing itself and all that we associate with it. We feel it too. Everything our attention rests upon triggers a sensation, most of which are minutely subtle and experienced beneath the level of conscious awareness. These feelings flicker and die so rapidly that they precede conscious thought, and thereby influence it. All these feelings reduce to just two impulses: advance and withdraw. As you scan any scene, then, you're in a storm of feeling; positive and negative sensations from the objects you see fall over you like fine drops of rain. This understanding is the beginning of creating a compelling and original character on the page. A character in fiction, like a character in life, inhabits their own unique hallucinated world in which everything they see and touch comes with its own unique personal meaning.

These worlds of feeling are a result of the way our brains

encode the environment. The models we have of everything are stored in the form of neural networks. When our attention rests upon a glass of red wine, say, a large number of neurons in different parts of the brain are simultaneously activated. We don't have a specific 'glass of wine' area that lights up, what we have are responses to 'liquid', 'red', 'shiny surface', 'transparent surface', and so on. When enough of these are triggered, the brain understands what's in front of it and constructs the glass of wine for us to 'see'.

But these neural activations aren't limited to mere descriptions of appearance. When we detect the glass of wine, other associations also flash into being: bitter-sweet flavours; vineyards; grapes; French culture; dark marks on white carpets; your road-trip to the Barossa Valley; the last time you got drunk and made a fool of yourself; the first time you got drunk and made a fool of yourself; the breath of the woman who attacked you. These associations have powerful effects on our perception. Research shows that when we drink wine our beliefs about its quality and price change our actual experience of its taste. The way food is described has a similar effect.

It's just such associative thinking that gives poetry its power. A successful poem plays on our associative networks as a harpist plays on strings. By the meticulous placing of a few simple words, they brush gently against deeply buried memories, emotions, joys and traumas, which are stored in the form of neural networks that light up as we read. In this way, poets ring out rich chords of meaning that resonate so profoundly we struggle to fully explain why they're moving us so.

Alice Walker's 'Burial' describes the poet bringing her child to the cemetery in Eatonton, Georgia, in which several generations of her family are interred. She describes her grandmother resting

> *undisturbed*
> *beneath the Georgia sun,*
> *above her the neatstepping hooves*
> *of cattle*

and graves that 'drop open without warning' and

> *cover themselves with wild ivy*
> *blackberries. Bittersweet and sage.*
> *No one knows why. No one asks.*

When I read 'Burial' for the first time, the lines at the end of this stanza made little logical sense to me, and yet I immediately found them beautiful, memorable and sad:

> *Forgetful of geographic resolutions as birds*
> *the far-flung young fly South to bury*
> *the old dead.*

It's these same associative processes that allow us to think metaphorically. Analyses of language reveal the extraordinary fact that we use around one metaphor for every ten seconds of speech or written word. If that sounds like too much, it's because you're so used to thinking metaphorically – to speaking of ideas that are 'conceived' or rain that is 'driving' or rage that is 'burning' or people who are 'dicks'. Our models are not only haunted by ourselves, then, but also by properties of other things. In her 1930 essay 'Street Haunting' Virginia Woolf employs several subtle metaphors over the course of a single gorgeous sentence:

> How beautiful a London street is then, with its islands of lights, and its long groves of darkness, and on the side of it perhaps some tree-sprinkled, grass-grown space where night is folding herself to sleep naturally and, as one passes the iron railing, one hears those little cracklings and stirrings of leaf and twig which seem to suppose the silence of fields all around them, an owl hooting, and far away the rattle of the train in the valley.

Neuroscientists are building a powerful case that metaphor is far more important to human cognition than has ever been imagined. Many argue it's the fundamental way that brains understand abstract concepts, such as love, joy, society and economy. It's simply not possible to comprehend these ideas in any useful sense, then, without attaching them to concepts that have physical properties: things that bloom and warm and stretch and shrink.

Metaphor (and its close sibling, the simile) tends to work on the page in one of two ways. Take this example, from Michael Cunningham's *A Home at the End of the World*: 'She washed old plastic bags and hung them on the line to dry, a string of thrifty tame jellyfish floating in the sun.' This metaphor works principally by opening an information gap. It asks the brain a question: how can a plastic bag be a jellyfish? To find the answer, we imagine the scene. Cunningham has nudged us into more vividly modelling his story.

In *Gone with the Wind*, Margaret Mitchell uses metaphor to make not a visual point, but a conceptual one: 'The very mystery of him excited her curiosity like a door that had neither lock nor key.'

In *The Big Sleep*, metaphor enables Raymond Chandler to pack a tonne of meaning into just seven words: 'Dead men are heavier than broken hearts.'

Brain scans illustrate the second, more powerful, use of metaphor. When participants in one study read the words 'he had a rough day', their neural regions involved in feeling textures became more activated, compared with those who read 'he had a bad day'. In another, those who read 'she shouldered the burden' had neural regions associated with bodily movement activated more than when they read 'she carried the burden'. This is prose writing that deploys the weapons of poetry. It works because it activates extra neural models that give the language additional meaning and sensation. We *feel* the heft and strain of the shouldering, we *touch* the abrasiveness of the day.

Such an effect is exploited by Graham Greene in *The Quiet American*. Here, a protagonist with a broken leg is receiving unwanted help from his antagonist: 'I tried to move away from him and take my own weight, but the pain came roaring back like a train in a tunnel.' This finely judged metaphor is enough to make you wince. You can almost feel the neural networks firing up and borrowing greedily from each other: the tender limb; the snapped bone; the pain in all its velocity and unstoppableness and thunder, roaring up the tunnel of the leg.

In *The God of Small Things*, Arundhati Roy uses metaphorical language to sensual effect when describing a love scene between the characters Ammu and Valutha: 'She could feel herself through him. Her skin. The way her body existed only where he touched her. The rest of her was smoke.'

And here the eighteenth-century writer and critic Denis Diderot uses a one-two of perfectly contrasting similes to smack his point home: 'Libertines are hideous spiders, that often catch pretty butterflies.'

Metaphor and simile can be used to create mood. In Karl Ove Knausgaard's *A Death in the Family*, the narrator describes stepping outside for a cigarette break, in the midst of clearing out the house of his recently deceased father. There he sees, 'plastic bottles lying on their sides on the brick floor dotted with raindrops. The bottlenecks reminded me of muzzles, as if they were small cannons with their barrels pointing in all directions.' Knausgaard's choice of language adds to the general deathly, angry aura of the passage by flicking unexpectedly at the reader's models of guns.

Descriptive masters such as Charles Dickens manage to hit our associative models again and again, creating wonderful crescendos of meaning, with the use of extended metaphors. Here he is, at the peak of his powers, introducing us to Ebenezer Scrooge in *A Christmas Carol*.

The cold within him froze his old features, nipped his pointed nose, shrivelled his cheek, stiffened his gait; made his eyes red,

his thin lips blue; and spoke out shrewdly in his grating voice.
A frosty rime was on his head, and on his eyebrows, and his
wiry chin. He carried his own low temperature always about
with him; he iced his office in the dog-days; and didn't thaw
it one degree at Christmas. External heat and cold had little
influence on Scrooge. No warmth could warm, nor wintry
weather chill him. No wind that blew was bitterer than he,
no falling snow was more intent upon its purpose, no pelting
rain less open to entreaty.

The author and journalist George Orwell knew the recipe for a potent metaphor. In the totalitarian milieu of his novel *Nineteen Eighty-Four*, he describes the small room in which the protagonist Winston and his partner Julia could be themselves without the state spying on them as 'a world, a pocket of the past where extinct animals could walk.'

It won't come as much of a surprise to discover the interminably correct Orwell was even right when he wrote about writing. 'A newly invented metaphor assists thought by evoking a visual image,' he suggested, in 1946, before warning against the use of that 'huge dump of worn-out metaphors which have lost all evocative power and are merely used because they save people the trouble of inventing phrases for themselves.'

Researchers recently tested this idea that clichéd metaphors become 'worn-out' by overuse. They scanned people reading sentences that included action-based metaphors ('they grasped the idea'), some of which were well-worn and others fresh. 'The more familiar the expression, the less it activated the motor system,' writes the neuroscientist Professor Benjamin Bergen. 'In other words, over their careers, metaphorical expressions come to be less and less vivid, less vibrant, at least as measured by how much they drive metaphorical simulations.'

1.8

In a classic 1932 experiment, the psychologist Frederic Bartlett read a traditional Native American story to participants and asked them to retell it, by memory, at various intervals. The War of the Ghosts was a brief, 330-word tale about a boy who was reluctantly compelled to join a war party. During the battle, a warrior warned the boy that he had been shot. But, looking down, the boy couldn't see any wounds on his body. The boy concluded that all the warriors were actually just ghosts. The next morning the boy's face contorted, something black came out of his mouth, and he dropped down dead.

The War of the Ghosts had various characteristics that were unusual, at least for the study's English participants. When they recalled the tale over time, Bartlett found their brains did something interesting. They simplified and formalised the story, making it more familiar by altering much of its 'surprising, jerky and inconsequential' qualities. They removed bits, added other bits and reordered still more. 'Whenever anything appeared incomprehensible, it was either omitted or explained,' in much the same way that an editor might fix a confusing story.

Turning the confusing and random into a comprehensible

story is an essential function of the storytelling brain. We're surrounded by a tumult of often chaotic information. In order to help us feel in control, brains radically simplify the world with narrative. Estimates vary, but it's believed the brain processes around 11 million bits of information at any given moment, but makes us consciously aware of no more than forty. The brain sorts through an abundance of information and decides what salient information to include in its stream of consciousness.

There's a chance you've been made aware of these processes when, in a crowded room, you've suddenly heard someone in a distant corner speaking your name. This experience suggests the brain's been monitoring myriad conversations and has decided to alert you to the one that might prove salient to your wellbeing. It's constructing your story for you: sifting through the confusion of information that surrounds you, and showing you only what counts. This use of narrative to simplify the complex is also true of memory. Human memory is 'episodic' (we tend to experience our messy pasts as a highly simplified sequences of causes and effects) and 'autobiographical' (those connected episodes are imbued with personal and moral meaning).

There's no single part of the brain that's responsible for such story making. While most areas have specialisms, brain activity is far more dispersed than scientists once thought. That said, we wouldn't be the storytellers we are if it wasn't for its most recently evolved region, the neocortex. It's a thin layer, about the depth of a shirt collar, folded in such a way that fully three feet of it is packed into a layer beneath your forehead. One of its critical jobs is keeping track of our social worlds. It helps interpret physical gestures, facial expressions and supports theory of mind.

But the neocortex is more than just a people-processor. It's also responsible for complex thought, including planning, reasoning and making lateral connections. When the psychologist Professor Timothy Wilson writes that one of the main differences between us and other animals is that we have a brain that's expert at constructing 'elaborate theories and explanations about what is happening in the world and why,' he's talking principally about the neocortex.

These theories and explanations often take the form of stories. One of the earliest we know of tells of a bear being chased by three hunters. The bear is hit. It bleeds over the leaves on the forest floor, leaving behind it all the colours of autumn, then manages to escape by climbing up a mountain and leaping into the sky, where it becomes the constellation Ursa Major. Versions of the 'Cosmic Hunt' myth have been found in Ancient Greece, northern Europe, Siberia, and in the Americas, where this particular one was told by the Iroquois Indians. Because of this pattern of spread, it's believed it was being told when there was a land bridge between what's now Alaska and Russia. That dates it between 13,000 and 28,000 BC.

The Cosmic Hunt myth reads like a classic piece of human bullshit. Perhaps it originated in a dream or shamanistic vision. But, just as likely, it started when someone, at some point, asked someone else, 'Hey, why do those stars look like a bear?' And that person gave a sage-like sigh, leaned on a branch and said, 'Well, it's funny you should ask . . .' And here we are, 20,000 years later, still telling it.

When posed with even the deepest questions about reality, human brains tend towards story. What is a modern religion if not an elaborate neocortical 'theory and explanation about what's happening in the world and why'? Religion doesn't merely seek to explain the origins of life, it's our answer to the most profound questions of all: What is good? What is evil? What do I do about all my love, guilt, hate, lust, envy, fear, mourning and rage? Does anybody love me? What happens when I die? The answers don't naturally emerge as data or an equation. Rather, they typically have a beginning, a middle and an end and feature characters with wills, some of them heroic, some villainous, all co-starring in a dramatic, changeful plot built from unexpected events that have meaning.

To understand the basis of how the brain turns the super-abundance of information that surrounds it into a simplified story is to understand a critical rule of storytelling. Brain stories have a basic structure of cause and effect. Whether it's memory, religion, or the War of the Ghosts, it rebuilds the confusion of reality into simplified theories of how one thing causes another. Cause and effect is a fundamental of how we understand the world. The brain can't help but make cause and effect connections. It's automatic. We can test it now. BANANAS. VOMIT. Here's the psychologist Professor Daniel Kahneman describing what just happened in your brain: 'There was no particular reason to do so, but your mind auto-matically assumed a temporal sequence and a causal connection between the words bananas and vomit, forming a sketchy scenario in which bananas caused the sickness.'

As Kahneman's test shows, the brain makes cause and effect connections even where there are none. The power of this cause and effect story-making was explored in the early twentieth century by the Soviet filmmakers Vsevolod Pudovkin and Lev Kuleshov, who juxtaposed film of a famous actor's expressionless face with stock footage of a bowl of soup, a dead woman in a coffin and a girl playing with a toy bear. They then showed each juxtaposition to an audience. 'The result was terrific,' recalled Pudovkin. 'The public raved about the acting of the artist. They pointed out the heavy pensiveness of his mood over the forgotten soup, were touched and moved by the deep sorrow with which he looked on the dead woman, and admired the light, happy smile with which he surveyed the girl at play. But we knew that in all three cases the face was exactly the same.'

Subsequent experiments confirmed the filmmakers' findings. When shown cartoons of simple moving shapes, viewers helplessly inferred animism and built cause-and-effect narratives about what was happening: this ball is bullying that one; this triangle is attacking this line, and so on. When presented with discs moving randomly on a screen, viewers imputed chase sequences where there were none.

Cause and effect is the natural language of the brain. It's how it understands and explains the world. Compelling stories are structured as chains of causes and effects. A secret of best-selling page-turners and blockbusting scripts is their relentless adherence to forward motion, one thing leading directly to another. In 2005, the Pulitzer prizewinning playwright David Mamet was captaining a TV drama called *The Unit*. After becoming frustrated with his writers producing scenes with no cause and effect – that were, for instance, simply there to deliver expository information – he sent out an angry ALL CAPS memo, which leaked online (I've de-capped what follows to save your ears): 'Any scene which does not both advance the plot and standalone (that is, dramatically, by itself, on its own merits) is either superfluous or incorrectly written,' he wrote. 'Start, every time, with this inviolable rule: the scene must be dramatic. It must start because the hero has a problem, and it must culminate with the hero finding him or herself either thwarted or educated that another way exists.'

The issue isn't simply that scenes without cause and effect tend to be boring. Plots that play too loose with cause and effect risk becoming confusing, because they're not speaking in the brain's language. This is what the screenwriter of *The Devil Wears Prada*, Aline Brosh McKenna, suggested when she said, 'You want all your scenes to have a "because" between them, and not an "and then".' Brains struggle with 'and then'. When one thing happens over here, and then we're with a woman in a car park who's just witnessed a stabbing, and then there's a rat in Mothercare in 1977, and then there's an old man singing sea shanties in a haunted pear orchard, the writer is asking a lot of people.

But sometimes this is on purpose. An essential difference between commercial and literary storytelling is its use of cause and effect. Change in mass-market story is quick and clear and easily understandable, while in high literature it's often slow and ambiguous and demands plenty of work from the reader, who has to ponder and de-code the connections for themselves. Novels such as Marcel Proust's *Swann's Way* are famously meandering and include, for example, a description of hawthorn blossom that lasts for well over a thousand words. ('You are fond of hawthorns,' one character remarks to the narrator, halfway through.) The art-house films of David Lynch are frequently referred to as 'dreamlike' because, like dreams, there's often a dearth of logic to their cause and effect.

Those who enjoy such stories are more likely to be expert readers, those lucky enough to have been born with the right kinds of minds, and raised in learning environments that nurtured the skill of picking up the relatively sparse clues in meaning left by such storytellers. I also suspect they tend to be higher than average in the personality trait 'openness to experience', which strongly predicts an interest in poetry and the arts (and also 'contact with psychiatric services'). Expert readers understand that the patterns of change they'll encounter in art-house films and literary or experimental fiction will be enigmatic and subtle, the causes and effects so ambiguous that they become a wonderful puzzle that stays with them months and even years after reading, ultimately becoming the source of meditation, re-analysis and debate with other readers and viewers – *why did characters behave as they did? What was the filmmaker really saying*?

But all storytellers, no matter who their intended audience, should beware of over-tightening their narratives. While it's dangerous to leave readers feeling confused and abandoned, it's just as risky to over-explain. Causes and effects should be *shown* rather than told; suggested rather than explained. Readers should be free to anticipate what's coming next and able to insert their own feelings and interpretations into why *that* just happened and what it all means. These gaps in explanation are the places in story in which readers insert themselves: their preconceptions; their values; their memories; their connections; their emotions – all become an active part of the story. No writer can ever transplant their neural world perfectly into a reader's mind. Rather, their two worlds mesh. Only by the reader insinuating themselves into a work can it create a resonance that has the power to shake them as only art can.

1.9

So our mystery is solved. We've discovered where a story begins: with a moment of unexpected change, or with the opening of an information gap, or likely both. As it happens to a protagonist, it happens to the reader or viewer. Our powers of attention switch on. We typically follow the consequences of the dramatic change as they ripple out from the start of the story in a pattern of causes and effects whose logic will be just ambiguous enough to keep us curious and engaged. But while this is technically true, it's actually only the shallowest of answers. There's obviously more to storytelling than this rather mechanical process.

A similar observation is made by a story-maker near the

start of Herman J. Mankiewicz and Orson Welles's 1941 cinema classic *Citizen Kane*. The film opens with change and an information gap: the recent death of the mogul Charles Foster Kane, as he drops a glass globe that contains a little snow-covered house and utters a single, mysterious word: *rosebud*. We're then presented with a newsreel that documents the raw facts of his seventy years of life: Kane was a well known yet controversial figure who was extraordinarily wealthy and once owned and edited the *New York Daily Inquirer*. His mother ran a boarding house and the family fortune came after a defaulting tenant left her a gold mine, the Colorado Lode, which had been assumed worthless. Kane was twice married, twice divorced, lost a son and made an unsuccessful attempt at entering politics, before dying a lonely death in his vast, unfinished and decaying palace that, we're told, was, 'since the pyramids, the costliest monument a man has built to himself'.

With the newsreel over, we meet its creators – a team of cigarette-smoking newsmen who, it turns out, have just finished their film and are showing it to their boss Rawlston for his editorial comments. And Rawlston is not satisfied. 'It isn't enough to tell us what a man did,' he tells his team. 'You've got to tell us who he was . . . How is he different from Ford? Or Hearst, for that matter? Or John Doe?'

That newsreel editor was right (as editors are with maddening regularity). We're a hyper-social species with domesticated brains that have been engineered specifically to control an environment of humans. We're insatiably inquisitive, beginning with our tens of thousands of childhood questions about how one thing causes another. Being a domesticated species, we're most interested of all in the cause and effect of other people. We're endlessly curious about them. What are they thinking? What are they plotting? Who do they love? Who do they hate? What are their secrets? What matters to them? Why does it matter? Are they an ally? Are they a threat? Why did they do that irrational, unpredictable, dangerous, incredible thing? What drove them to build 'the world's largest pleasure ground' on top of a manmade 'private mountain' that contained the most populous zoo 'since Noah' and a 'collection of everything so big it can never be catalogued'? Who is the person really? How did they become who they are?

Good stories are explorations of the human condition; thrilling voyages into foreign minds. They're not so much about events that take place on the surface of the drama as they are about the characters that have to battle them. Those characters, when we meet them on page one, are never perfect. What arouses our curiosity about them, and provides them with a dramatic battle to fight, is not their achievements or their winning smile. It's their flaws.

CHAPTER TWO:

THE FLAWED SELF

GRIFFITH COLLEGE DUBLIN
SOUTH CIRCULAR ROAD DUBLIN 8.
Tel: 01 4150490 Fax: (01) 4549265
library@griffith.ie

GRIFFITH COLLEGE DUBLIN
SOUTH CIRCULAR ROAD, DUBLIN 8
Tel. 01 4150400 Fax (01) 4549265
www.gcd.ie

2.0

There's something you should know about Mr B. He's being watched by the FBI. They film him constantly and in secret, then cut the footage together and broadcast it to millions as 'The Mr B Show'. This makes life rather awkward for Mr B. He showers in swimming trunks and dresses beneath bedsheets. He hates talking to others, as he knows they're actors hired by the FBI to create drama. How can he trust them? He can't trust anyone. No matter how many people explain why he's wrong, he just can't see it. He finds a way to dismiss each argument they present to him. He *knows* it's true. He *feels* it's true. He sees evidence for it *everywhere*.

There's something else you should know about Mr B. He's psychotic. One healthy part of his brain, writes the neuro-scientist Professor Michael Gazzaniga, 'is trying to make sense out of some abnormalities going on in another'. The malfunctioning part is causing 'a conscious experience with very different contents than would normally be there, yet those contents are what constitute Mr B's reality and provide experiences that his cognition must make sense of.'

Because it's being warped by faulty signals being sent out by the unhealthy section of his brain, the story Mr B is telling about the world, and his place within it, is badly mistaken. It's so mistaken he's no longer able to adequately control his environment, so doctors and care staff have to do it on his behalf, in a psychiatric institution.

As unwell as he is, we're all a bit like Mr B. The controlled hallucination inside the silent, black vault of our skulls that we experience as reality is warped by faulty information. But because this distorted reality is the only reality we know, we just can't see where it's gone wrong. When people plead with us that we're mistaken or cruel and acting irrationally, we feel driven to find a way to dismiss each argument they present to us. We *know* we're right. We *feel* we're right. We see evidence for it *everywhere*.

These distortions in our cognition make us flawed. Everyone is flawed in their own interesting and individual ways. Our flaws make us who we are, helping to define our character. But our flaws also impair our ability to control the world. They harm us.

At the start of a story, we'll often meet a protagonist who is flawed in some closely defined way. The mistakes they're making about the world will help us empathise with them. We'll warm to their vulnerability. We'll become emotionally engaged in their struggle. When the dramatic events of the plot coax them to change we'll root for them.

The problem is, in fiction and in life, changing who we are is hard. The insights we've learned from neuroscience and psychology begin to show us exactly *why* it's hard. Our flaws – especially the mistakes we make about the human world and how to live successfully within it – are not simply ideas about this and that which we can identify easily and choose to shrug off. They're built right into our hallucinated models. Our flaws form part of our perception, our experience of reality. This makes them largely invisible to us.

Correcting our flaws means, first of all, managing the task of actually seeing them. When challenged, we often respond by refusing to accept our flaws exist at all. People accuse us of being 'in denial'. Of course we are: we literally can't see them. When we *can* see them, they all too often appear not as flaws at all, but as virtues. The mythologist Joseph Campbell identified a common plot moment in which protagonists 'refuse the call' of the story. This is often why.

Identifying and accepting our flaws, and then changing who we are, means breaking down the *very structure of our reality* before rebuilding it in a new and improved form. This is not easy. It's painful and disturbing. We'll often fight with all we have to resist this kind of profound change. This is why we call those who manage it 'heroes'.

There are various routes by which characters and selves become unique and uniquely flawed, and a basic under-standing of them can be of great value to storytellers. One major route involves those moments of change. The brain constructs its hallucinated model of the world by observing millions of instances of cause and effect then constructing its own theories and assumptions about how one thing caused the other. These micro-narratives of cause and effect – more commonly known as 'beliefs' – are the building blocks of our neural realm. The beliefs it's built from feel personal to us because they help make up the world that we inhabit and our understanding of who we are. Our beliefs feel personal to us because they *are* us.

But many of them will be wrong. Of course the controlled hallucination we live inside is not as distorted as the one that Mr B lives inside. Nobody, however, is right about everything. Nevertheless, the storytelling brain wants to sell us the illusion that we are. Think about the people closest to you. There won't be a soul among them with whom you've never disagreed. You know *she's* slightly wrong about that, and *he's* got that wrong, and don't get *her* started on that. The further you travel from those you admire, the more wrong people become until the only conclusion you're left with is that entire tranches of the human population are stupid, evil or insane. Which leaves you, the single living human who's right about everything – the perfect point of light, clarity and genius who burns with godlike luminescence at the centre of the universe.

Hang on, that can't be right. You must be wrong about *something*. So you go on a hunt. You count off your most precious beliefs – the ones that really matter to you – one by one. You're not wrong about *that* and you're not wrong about *that* and you're certainly not wrong about *that* or *that* or *that* or *that*. The insidious thing about your biases, errors and prejudices is that they appear as real to you as Mr B's delusions appear to him. It feels as if everyone else is 'biased' and it's only you that sees reality as it actually is. Psychologists call this 'naive realism'. Because reality seems clear and obvious and self-evident to you, those who claim to see it differently must be idiots or lying or morally derelict. The characters we meet at the start of story are, like most of us, living just like this – in a state of profound naivety about how partial and warped their hallucination of reality has become. They're wrong. They don't know they're wrong. But they're about to find out . . .

If we're all a bit like Mr B then Mr B is, in turn, like the protagonist in Andrew Niccol's screenplay, *The Truman Show*. It tells of thirty-year-old Truman Burbank, who's come to believe his whole life is staged and controlled. But, unlike Mr B, he's right. *The Truman Show* is not only real, it's being broadcast, twenty-four hours a day, to millions. At one point, the show's executive producer is asked why he thinks it's taken Truman so long to become suspicious of the true nature of his world. 'We accept the reality of the world with which we're presented,' he answers. 'It's as simple as that.'

We certainly do. As wrong as we are, we rarely question the reality our brains conjure for us. It is, after all, our 'reality'. As well as this, the hallucination is functional. Each one of the tiny beliefs that make up our neural model is a little instruction that tells our brain how the outside world works: *this is how you open a stuck jam jar lid; this is how you lie to a police officer; this is how you behave if you want your boss to believe you're a useful, sane and honest employee.* These instructions make our environment predictable. They make it controllable. Taken in sum, the vastly intricate web of beliefs can be seen as the brain's 'theory of control'. It's this theory of control that's often challenged at the story's start.

In his celebrated novel *The Remains of the Day*, Kazuo Ishiguro takes us into the warped and flawed neural realm of James Stevens, a proud head butler in a large stately home. We learn that his core beliefs about the world and how to control it came from his father, Stevens Senior, who was a butler of prodigious talent. Stevens is passionate about his calling and muses about the 'special quality' that made his father, and butlers like him, so great. 'Dignity', he decides, the key to which is 'emotional restraint'. Just as the English landscape is beautiful because of its 'lack of obvious drama or spectacle', a great butler 'will not be shaken out by external events, however surprising, alarming or vexing'.

Emotional restraint is why the English make the best butlers. 'Continentals are unable to be butlers because they are as a breed incapable of the emotional restraint which only the English race are capable of.' They, and the Celts for that matter, 'are like a man who will, at the slightest provocation, tear off his suit and his shirt and run about screaming.' Emotional restraint is the pivotal idea around which his neural model of the world is built. It's his theory of control. If he adheres to it, he'll be able to manipulate his environment in such a way that he'll get what he wants, namely, the reputation of a brilliant butler. This flawed belief defines him. It *is* him. It's characters like Stevens, who inhabit their flaw with such concentrated precision, that often prove to be the most memorable, immediate and compelling.

Ishiguro's book softly yet brutally exposes the ways in which Steven's flawed perceptions of reality have harmed him. Its most crushing scenes play out one evening, as Stevens is captaining an important function at the house. Upstairs, his elderly father, finally broken by a lifetime of service, has just come around after suffering a collapse. A preoccupied Stevens is persuaded to see him. Perhaps sensing the gravity of his situation, Stevens Senior breaks through his own ironclad armour of emotional restraint and expresses a hope that he's been a good father. His son can only respond with an awkward laugh. 'I'm so glad you're feeling better now,' he says. His father tells him he's proud of him. Then he pushes the point, 'I hope I've been a good father to you. I suppose I haven't.'

'I'm afraid we're extremely busy now,' his son replies. 'But we can talk again in the morning.'

Later that evening, Stevens Senior has a stroke. He's on the edge of death. His son is coaxed up to see him again and, again, insists he must return to his duties. Downstairs his boss, Lord Darlington, senses something's wrong. 'You look as though you're crying,' he says. Stevens quickly dabs the corners of his eyes and laughs, 'I'm very sorry, sir. The strains of a hard day.' When his father dies, shortly afterwards, Stevens is again too busy to attend. 'I know my father would have wished me to carry on just now,' he remarks to a maid. And there's little doubt he's correct.

The brilliance of this sequence – its psychological truth – is that this is not a memory of shame and regret, for Stevens, but one of victory. In fact, it's his pitch for being held in the pantheon of the Britain's greatest and most dignified butlers. 'For all its sad associations,' he says, 'whenever I recall that evening today, I find I do so with a large sense of triumph.' The hallucinated model Stevens had of reality was built around the value of emotional restraint. That was the core of his brain's theory about how a person should control the world. And, as far as he was concerned, he'd aced it.

Stevens's neural world was warped and twisted and yet, just like Mr B, he saw evidence all around him that it was entirely accurate. After all, hadn't his model of reality and its theory of control worked? Hadn't his belief in the sacred value of emotional restraint given him his career, his status and protected him from the pain of losing his father? Ishiguro's novel is an exploration of the truth of that flaw and its ramifications – how, as Salman Rushdie has written, Stevens was, 'destroyed by the ideas upon which he has built his life'.

The mythologist Joseph Campbell said that 'the only way you can describe a human being truly is by describing his imperfections.' It's this imperfect person we meet in story and in life. But unlike in life, story allows us to crawl into that character's mind and understand them. For us hyper-social domesticated creatures, there's little more fascinating than the cause and effect of other people, the 'why' of what people do as they do. But story offers more than just this. Locked inside the black vault of our skulls, stuck forever in the solitude of our own hallucinated universe, story is a portal, a hallucination within the hallucination, the closest we'll ever really come to escape.

2.1

When designing a character, it's often useful to think of them in terms of their theory of control. How have they learned to control the world? When unexpected change strikes, what's their automatic go-to tactic for wrestling with the chaos? What's their default, flawed response? The answer, as we've just seen, comes from that character's core beliefs about reality, the precious and fiercely defended ideas around which they've formed their sense of self.

But who we are, in all our partiality and weirdness, is also partly genetic. Our genes begin to guide the way our brains and hormonal systems are wired up when we're in the womb. We enter the world semi-finished. Then, early life events and influences work in combination with genes to build our core personality. Unless something terrible happens to psychologically break us, this personality is likely to remain relatively stable throughout our life, changing only modestly and in predictable ways as we age.

Psychologists measure personality across five domains, which can be useful for writers doing character work to know. Those high in extraversion are gregarious and assertive, seekers of attention and sensation. Being high in neuroticism means you're anxious, self-conscious and prone to depression, anger and low self-esteem. Lots of openness makes for a curious soul, someone artistic, emotional and comfortable with novelty. High-agreeable people are modest, sympathetic and trusting while their disagreeable opposites have a competitive and aggressive bent. Conscientious people prefer order and discipline and value hard work, duty and hierarchy. Psychologists have applied these domains to fictional characters. One academic paper included the following examples:

Neuroticism (high): Miss Havisham (*Great Expectations*, Charles Dickens)

Neuroticism (low): James Bond (*Casino Royale*, Ian Fleming)

Extraversion (high): The Wife of Bath (*The Canterbury Tales*, Geoffrey Chaucer)

Extraversion (low): Boo Radley (*To Kill a Mockingbird*, Harper Lee)

Openness (high): Lisa Simpson (*The Simpsons*, Matt Groening)

Openness (low): Tom Buchanan (*The Great Gatsby*, F. Scott Fitzgerald)

Agreeableness (high): Alexei Karamazov (*The Brothers Karamazov*, Fyodor Dostoyevsky)

Agreeableness (low): Heathcliff (*Wuthering Heights*, Emily Brontë)

Conscientiousness (high): Antigone (*Antigone*, Sophocles)

Conscientiousness (low): Ignatius J. Reilly (*A Confederacy of Duncers*, John Kennedy Toole)

These 'big five' personality traits aren't switches – we're not one thing or the other. Rather, they're dials, with us having more or less of each trait, our particular highs and lows combining to form our own peculiar self. Personality has a powerful influence over our theory of control. Different personalities have different go-to tactics for controlling the environment of people. When unexpected change threatens, some are more likely to jump to aggression and violence, some charm, some flirtation, others will argue or withdraw or become infantile or try to negotiate for consensus or become Machiavellian or dishonest, resorting to threat, bribery or con.

This, then, is how unique and interesting fictional characters generate unique and interesting plots. 'It is from character,' writes the psychologist Professor Keith Oatley, 'that flow goals, plans and actions.' As we interact with the world in our own characteristic way, so the world pushes back in ways which reflect it, setting us off in our own particular cause-and-effect journey – a plot specific to us. A disagreeable neurotic sending out grumpy, twitchy causes into the world has to deal with the negative effects that fly back. A feedback loop of grumpiness emerges, with the neurotic convinced they're behaving reasonably and rationally only to be tossed, once again, into an oubliette of hostility and disapproval. One extra episode of paranoia or irritation per week will trigger enough negativity in other people that they'll find themselves living in a neural realm that's entirely different from the average smiley high-agreeable. It's in these ways that tiny differences in brain structure can add up to massively different lives and plots.

Personality can predict what kinds of futures we might have too. Conscientious people tend to enjoy greater than average job security and life satisfaction; extroverts are more likely to have affairs and car accidents; disagreeable people are better at fighting their way up corporate ladders into the highest-paying jobs; those high in openness are more likely to get tattoos, be unhealthy and vote for left-wing political parties while those low in conscientiousness are more likely to end up in prison and have a higher risk of dying, in any given year, of around 30 per cent. Although women and men are far more alike than they are different, there are gender differences. One of the most reliable findings in the literature is that males tend to be more disagreeable than females, with the average man scoring lower in agreeability than around 60 per cent (and, in some studies, 70 per cent) of women. A similar personality gap is found for neuroticism, where the average man scores lower than around 65 per cent of women.

As a person low in extraversion and high in neuroticism, writing to you from the corner of a darkened room in a cottage that lies at the end of a crumbling path, deep in the Kent countryside, I can attest to the extent to which traits can guide fates. The butler Stevens would've been attracted to his life of service in part because of his personality, which seems unusually high in conscientiousness and low in openness and extraversion. He'd have inherited these traits from his much-admired father because personality, of course, is significantly heritable. Charles Foster 'Citizen' Kane, meanwhile, was low in agreeableness, low in neuroticism and high in extraversion: he was monstrously ambitious, lacked self-doubt and craved the approval of others. It was these three qualities, more than any others, that defined his personality and dictated the decisions which formed the plot of his life.

2.2

Storytellers can show the personality of their characters in almost everything they do: it's in their thoughts, dialogue, social behaviours, memories, desires and sadnesses. It's in how they behave in traffic jams, what they think of Christmas and their reaction to a bee. 'Human personalities are rather like fractals,' writes the psychologist Professor Daniel Nettle.

'It is not just that what we do in the large-scale narratives of our lives – love, career, friendships – tends to be somewhat consistent over time, with us often repeating the same kinds of triumph or mistakes. Rather, what we do in tiny interactions like the way we shop, dress or talk to a stranger on the train or decorate our houses, shows the same kinds of patterns as can be observed from examining a whole life.'

Human environments are rich with clues about those who occupy them. People make 'identity claims' to broadcast who they are. This could be through displaying certificates, books, tattoos or meaningful objects. Identity claims betray how these people want others to think of them. People use 'feeling regulators', motivational posters, scented candles or items that make them feel nostalgic, excited or loved. Extroverts who feel energised by bright colours are more likely to decorate their homes or dress accordingly, while introverts prefer the hush of muted tones. 'Behavioural residue' is what psychologists call the things we accidentally leave behind: the stashed wine bottle, the torn-up manuscript, the punch dent in the wall. The psychologist Professor Sam Gosling advises the curious to 'look out for discrepancies in the signals that people send to themselves and others'. Broadcasting one version of self in their private spaces and another in their hallways, kitchens and offices can hint at a tortuous 'fractionating of the self'.

In her novel *Notes on a Scandal*, Zoë Heller makes brilliant use of home environments to feed our neural models of its two central characters. When the narrator Barbara Covett (low in openness and agreeableness, high in conscientiousness) visits the home of Sheba Hart (the opposite) we're treated to a rich insight into their contrary personalities. Covett recalls that, on the rare occasion she has visitors to her flat she cleans it 'scrupulously' and even grooms the cat. And yet she still experiences 'the most terrible feeling of exposure . . . as if my dirty linen, rather than my unexceptional sitting room, were on display'. Not so Sheba. When Barbara enters her living room she sees in it a 'bourgeois confidence' and a 'level of disorder . . . I could never tolerate'. There is 'tatty, gigantic furniture', 'her children's stray underwear', 'a primitive wooden instrument, possibly African, which looked as if it might be rather smelly'. The mantelpiece is 'a gathering point for household flotsam. A child's drawing. A hunk of pink Play-Doh. A passport. One elderly-looking banana.'

The environment triggers, in Barbara, a reaction that surprises her: the clutter makes her envious. This, in turn, sparks a melancholy thought that illuminates her character further and also relies on the way personality helplessly leaks into the spaces we occupy.

When you live alone, your furnishings, your possessions, are always confronting you with the thinness of your existence. You know with painful accuracy the provenance of everything you touch and the last time you touched it. The five little cushions on your sofa stay plumped and leaning at their jaunty angle for months at a time unless you theatrically muss them. The level of the salt in your shaker decreases at the same excruciating rate, day after day. Sitting in Sheba's house – studying the mingled detritus of

*its several inhabitants – I could see what a relief it might be
to let your own meagre effects be joined with other people's.*

In this vivid and touching passage we hear the howl of the
lonely in five plumped cushions and salt.

Our habit of leaving revealing clues in our environment is
why journalists prefer interviewing subjects in their homes.
When Lynn Barber met the formidable architect Zaha Hadid,
she was let into her 'bare white penthouse' by a publicist prior
to Hadid's arrival. The flat, in which she'd lived for two and a
half years, had 'all the intimacy of a car showroom', wrote
Barber.

*It is extremely, dauntingly, hard. There are no curtains,
carpets, cushions or upholstery of any kind. The furniture,
if that's the right word, consists of slippery amorphous
shapes made of reinforced fibreglass and painted with car
paint . . . Her bedroom is fractionally more inviting in that
it does at least have a recognisable bed, a small oriental rug,
and a table with all her jewellery and scent bottles laid out,
but that's about it.'*

Rooms, she wrote, 'are supposed to provide clues to personality, but this seems to be a statement of impersonality'. Of
course, Barber's vivid and telling descriptions richly fed our
models of Hadid's mind. We began to know who she was
before she'd even walked in.

2.3

As powerful a force as personality is, we're more than just introverts, extraverts and the rest. Our traits work with our cultural, social and economic environments, as well as the experiences we go through, to construct a neural world for us to live in that is unique.

There's little more thrilling, in a story, than suddenly encountering a mind that is utterly different to ours while being revealing of character and the story to come. The protagonist's point of view orients us in the story. It's a map of clues, full of hints about its owner's flaws and the plot they're going to create. For me, it's the single most underrated quality of fiction writing. Too many books and films begin with characters that seem to be mere outlines: perfect, innocent human-shaped nothings, perhaps with a bolt-on quirk or two, waiting to be coloured in by the events of the plot. Far better to find ourselves waking up, on page one, startled and exhilarated to find ourselves inside a mind and a life that feels flawed, fascinating, specific and real.

Charles Bukowski manages this brilliantly in the opening paragraph of his novel *Post Office*:

> *It began as a mistake.*
> *It was Christmas season and I learned from the drunk up on the hill, who did the trick every Christmas, that they would hire damned near anybody, and so I went and the next thing I knew I had this leather sack on my back and was hiking*

around at my leisure. What a job, I thought. Soft! They only gave you a block or two and if you managed to finish, the regular carrier would give you another block to carry, or maybe you'd go back in and the soup would give you another, but you just took your time and shoved those Xmas cards in the slots.

A world away from blue-collar Los Angeles, Zadie Smith's *White Teeth* opens in Cricklewood Broadway at the scene of the attempted suicide of 47-year-old Archie Jones, 'dressed in corduroy and sat in a fume-filled Cavalier Musketeer Estate . . . scrunched up in each fist he held his army service medals (left) and his marriage licence (right), for he had decided to take his mistakes with him . . . He wasn't the type to make elaborate plans – suicide notes and funeral instructions – he wasn't the type for anything fancy. All he asked for was a bit of silence, a bit of shush so he could concentrate . . . He wanted to do it before the shops opened.'

In most of the best contemporary fiction, objects and events aren't usually described from a God-like view, but from the unique perspective of the character. As in life, everything we encounter is a component not of objective external reality, but of that character's inner neural realm – the controlled hallucination that, no matter how real it seems, exists only in their head and is, in its own way, wrong. In fiction, it might not be going too far to say *all* description works as a description of character.

In an electrifying passage from his novel *Another Country*, James Baldwin shows Rufus Scott – a doomed African-

American trying to survive in 1950s America – walking into a Harlem jazz club. Baldwin's description of the saxophonist playing on the stage crackles with as much information about Scott, his world and his frustrated attempts at controlling it, as it does about the musician, who he perceives,

> wide-legged, humping the air, filling his barrel chest, shivering in the rags of his twenty-odd years, and screaming through the horn Do you love me? Do you love me? Do you love me? And again Do you love me? Do you love me? Do you love me? This, anyway, was the question Rufus heard, the same phrase, unbearably endlessly, and variously repeated with all the boy had . . . the question was terrible and real; the boy was blowing with his lungs and guts out of his own short past; somewhere in that past, in the gutters or gang fights or gang shags; in the acrid room, on the sperm-stiffened blanket, behind marijuana or the needle, under the smell of piss in the precinct basement, he had received the blow from which he never would recover and this no one wanted to believe. Do you love me? Do you love me? Do you love me?

2.4

Culture is another route by which characters in life and fiction become the flawed and peculiar people they are. We often think of 'culture' as surface phenomena, such as opera and literature and modes of dress, but culture is actually built deeply and directly into our model of the world. It forms part of the neural machinery that constructs our hallucination of reality. Culture distorts and narrows the lens through which we experience life, exerting a potent influence us, whether by dictating the moral rules we'll fight and die to defend or defining the kinds of foods we'll perceive as delicious. The Japanese eat *hachinoko*, a delicacy made from baby bees. The Korowai of Papua New Guinea eat people. Americans consume ten billion kilograms of beef a year, while in India, where cows are sacred, a vigilante might kill you for eating a steak sandwich. Orthodox Jewish wives shave their heads and wear wigs, lest any alluring trace of hair be glimpsed by dirty mortals. The Waorani of Ecuador wear almost nothing at all.

Such cultural norms are incorporated into our models in childhood, a period in which the brain is rapidly working out who it needs to be in order to best control its particular environment. Between the ages of zero and two, it generates around 1.8 million neural connections every second. It remains in this state of increased malleability – or 'plasticity' – until late adolescence or early adulthood. It learns, in part, through playing. Lots of animals enjoy these pleasurable, rule-based, exploratory interactions, including dolphins, kangaroos and rats. But our domestication, and the highly complex social realm we must learn to control, has elevated the importance of play in humans. It's the main reason we have such greatly extended childhoods.

We've evolved different forms of play, from games to education to storytelling. Play, including storytelling, is typically overseen by adults who tell children what's fair and not fair, what's of value and not, and how we should behave, punishing and rewarding when we act in accordance, or not, to the models of our culture. Caregivers don't merely read morally charged stories to their children, they often add their own narration, underlining the narrative's message. Play is critical for the making of social minds. One study into the backgrounds of sociopathic murderers found no connection between them apart from an extreme lack of play, or a history of abnormal play such as sadism and bullying, in the childhoods of 90 per cent of them.

It's in our first seven years that culture mostly gets built into our models, honing and particularising our neural realm. Western children are raised in a culture of individualism which was birthed around 2,500 years ago in Ancient Greece. Individualists tend to fetishise personal freedom and perceive the world as being made up of individual pieces and parts. This gives us a set of particular values that strongly influence the stories we tell. According to some psychologists, it's a mode of thinking that arose from the physical landscape of Ancient Greece. It was a rocky, hilly, coastal place, and therefore poor for large group endeavours like farming. This meant you had to be something of a hustler to get by – a small business person tanning hides, perhaps, or foraging or making olive oil or fishing. The best way of controlling that world, in Ancient Greece, was by being self-reliant.

Because individual self-reliance was the key to success, the all-powerful individual became a cultural ideal. The Greeks sought personal glory and perfection and fame. They created that legendary competition of self versus self, the Olympics, practised democracy for fifty years and became so self-focused they felt compelled to warn of the dangers of runaway self-love in the story of Narcissus. This conception of the individual as the locus of their own power, free to choose the life they wanted, rather than being slave to the whims of tyrants, fates and gods, was revolutionary. It 'changed the way people thought about cause and effect,' writes the psychologist Professor Victor Stretcher, 'heralding in Western civilisation'.

Compare this pushy, freedom-loving self to the one that emerged in the East. The undulating and fertile landscape in Ancient China was perfect for large groupish endeavours. Getting by would have probably meant being a part of a sizeable wheat- or rice-growing community or working on a huge irrigation project. The best way of controlling the world, in that place, was ensuring the group, rather than the individual, was successful. That meant keeping your head down and being a team player. This collective theory of control led to a collective ideal of self. In the Analects, Confucius is recorded as describing 'the superior man' as one who 'does not boast of himself", preferring instead the 'concealment of his virtue'. He 'cultivates a friendly harmony' and 'lets the states of equilibrium and harmony exist in perfection'. He could hardly be more different than the pushy Westerner emerging seven thousand kilometres away.

For the Greeks, the primary agent of control was the individual. For the Chinese, it was the group. For the Greeks, reality was made up of individual pieces and parts. For the Chinese, it was a field of interconnected forces. Out of these differences in the experience of reality come different story forms. Greek myths usually have three acts, Aristotle's 'beginning, middle and end', perhaps more usefully described as crisis, struggle, resolution. They often starred singular heroes battling terrible monsters and returning home with treasures.

This was individualist propaganda, transmitting the notion that one courageous person really could change everything. These story outlines begin influencing a Western child's emerging self surprisingly early. On being asked by researchers to spontaneously tell a story, one three-year-old girl in the US produced a perfect sequence of crisis-struggle-resolution: 'Batman went away from his mommy. Mommy said, "Come back, come back." He lost and his mommy can't find him. He ran like this to come home. He eat muffins and he sat on his mommy's lap. And then him have a rest.'

Stories weren't like this in Ancient China. This was a realm so other-focused there was practically no real autobiography for two thousand years. When it did finally emerge, life stories were typically told stripped of the subject's voice and opinions and they were positioned not at the centre of their own lives but as a bystander looking in. Rather than following a straightforward pattern of cause and effect, Eastern fiction often took the form of Ryūnosuke Akutagawa's 'In A Bamboo Grove', in which the events surrounding a murder are recounted from the perspectives of several witnesses – a woodcutter, a priest, a policeman, an elderly woman, the accused murderer, the victim's wife, and finally from a spirit medium channelling the victim himself. All these accounts somehow contradict each other, with the reader left to puzzle out their meaning for themselves.

In such stories, according to the psychologist Professor

Uichol Kim, 'you're never given the answer. There's no closure. There's no happily ever after. You're left with a question that you have to decide for yourself. That's the story's pleasure.' In Eastern tales that did focus on an individual, the hero's status tended to be earned in a suitably group-first way. 'In the West you fight against evil and the truth prevails and love conquers all,' he said. 'In Asia it's a person who sacrifices who becomes the hero, and takes care of the family and the community and the country.'

The Japanese form known as Kishõtenketsu comes with four acts: in act one ('ki') we're introduced to the characters, in act two ('sho') the actions follow on, in act three ('ten') a twist that's surprising or even apparently unconnected takes place and in the final act ('ketsu') we're invited, in some open-ended way, to search for the harmony between it all. 'One of the confusing things about stories in the East is there's no ending,' said Professor Kim. 'In life there are not simple, clear answers. You have to find these answers.'

Whereas Westerners enjoy having accounts of individual struggle and victory beamed into their neural realms, Easterners take pleasure from the narrative pursuit of harmony.

What these forms reflect is the different ways our cultures understand change. For Westerners, reality is made up of individual pieces and parts. When threatening unexpected change strikes, we tend to reimpose control by going to war with those pieces and parts and trying to tame them. For Easterners, reality is a field of interconnected forces. When threatening unexpected change strikes, they're more likely to reimpose control by attempting to understand how to bring those turbulent forces back into harmony so that they can all exist together. What they have in common is story's deepest purpose. They teach lessons in control.

2.5

It takes time for a self, with all its flaws and peculiarities, to bend itself out of the universe. It begins with us recognising our image in the mirror. Our caregivers tell us stories about the past and the present, what's happening around us and what we had to do with it. We begin to contribute to these little stories about ourselves. We realise we're goal-directed – we want things and we try to get them. We grasp that we're surrounded by other minds that are also goal-directed. We understand ourselves to be a certain category of human – a girl, a boy, working-class – of whom others have specific expectations. We have power and have done things. These pockets of story memory slowly begin to connect and cohere. They form plots that become imbued with character and theme. Finally, in adolescence, writes the psychologist Professor Dan McAdams, we endeavour to understand our life as a 'grand narrative, reconstructing the past and imagining the future in such a way as to provide it with some semblance of purpose, unity and meaning'.

Having undergone its adolescent narrative-making process, the brain has essentially worked out who we are, what matters, and how we should behave in order to get what we want. Since birth, it's been in a state of heightened plasticity that has enabled it to build its models. But now it becomes less plastic and harder to change. Most of the peculiarities and mistakes that make us who we are have become incorporated into its models. Our flaws and peculiarities have become who we are. Our minds have been made up.

Then the brain enters a state that's valuable to understand for anyone interested in human conflict and drama. From being model-builders we become model defenders. Now that the flawed self with its flawed model of the world has been constructed, the brain starts to protect it. When we encounter evidence that it might be wrong, because other people aren't perceiving the world as we do, we can find it deeply disturbing. Rather than changing its models by acknowledging the perspectives of these people, our brains seek to deny them.

This is how the neurobiologist Professor Bruce Wexler describes it: 'Once [the brain's] internal structures are established they turn the relationship between the internal and external around. Instead of the internal structures being shaped by the environment, the individual now acts to preserve established structures in the face of environmental challenges, and finds changes in structure difficult and painful.' We respond to such challenges with distorted thinking, argument and aggression. As Wexler writes, 'we ignore, forget or attempt to actively discredit information that is inconsistent with these structures'.

The brain defends our flawed model of the world with an armoury of crafty biases. When we come across any new fact or opinion, we immediately judge it. If it's consistent with our model of reality our brain gives a subconscious feeling of *yes*.

If it's not, it gives a subconscious feeling of *no*. These emotional responses happen before we go through any process of conscious reasoning. They exert a powerful influence over us. When deciding whether to believe something or not, we don't usually make an even-handed search for evidence. Instead, we hunt for any reason to confirm what our models have instantaneously decided for us. As soon as we find any half-decent evidence to back up our 'hunch' we think, 'Yep, that makes sense.' And then we stop thinking. This is sometimes known as the 'makes sense stopping rule'.

Not only do our neural-reward systems spike pleasurably when we deceive ourselves like this, we kid ourselves that this one-sided hunt for confirmatory information was noble and thorough. This process is extremely cunning. It's not simply that we ignore or forget evidence that goes against what our models tell us (although we do that too). We find dubious ways of rejecting the authority of opposing experts, give arbitrary weight to some parts of their testimony and not others, lock onto the tiniest genuine flaws in their argument and use them to dismiss them entirely. Intelligence isn't effective at dissolving these cognitive mirages of rightness. Smart people are mostly better at finding ways to 'prove' they're right and tend to be no better at detecting their wrongness.

It might seem odd that humans have evolved to be so irrational. One compelling theory has it that, because we evolved in groups, we're designed to argue things out lawyer-style until the optimal way forward emerges. Truth, then, is a group activity and free speech an essential component. This would validate the screenwriter Russell T. Davies's observation that good dialogue is 'two monologues clashing. It's true in life, never mind drama. Everyone is always, always thinking about themselves.'

Because our models make up our actual experience of reality, it's little wonder that any evidence which suggests they are wrong is profoundly unsettling. 'Things are experienced as pleasurable because they are familiar,' writes Wexler, 'while the loss of the familiar produces stress, unhappiness and dysfunction.' We're so used to our aggressive model-defending responses – they're such an ordinary part of being alive – we become inured to their strangeness. Why do we dislike people we disagree with? Why do we feel emotionally repulsed by them?

The rational response, when encountering someone with alien ideas, would be to either attempt to understand them or shrug. And yet we become distressed. Our threatened neural models generate waves of sometimes overwhelming negative feelings. Incredibly, the brain treats threats to our neural models in the much same way as it defends our bodies from a physical attack, putting us into a tense and stressful fight-or-flight state. The person with merely differing views becomes a dangerous antagonist, a force that's actively attempting to harm us. The neuroscientist Professor Sarah Gimbel watched what happened when people in brain scanners were presented with evidence their strongly held political beliefs were wrong. 'The response in the brain that we see is very similar to what would happen if, say, you were walking through the forest and came across a bear,' she has said.

So we fight back. We might do so by trying to convince our opponent of their wrongness and our rightness. When we fail, as we usually do, we can be thrown into torment. We chew the conflict over and over, as our panicked mind lists more and more reasons why they're dumb, dishonest or morally corrupt. Indeed, language provides a stinking rainbow of words for people whose mental models conflict with ours: idiot, cretin, imbecile, pillock, berk, arsehole, airhead, sucker, putz, barnshoot, crisp-packet, clown, dick, divot, wazzock, fuckwit, fucknut, titbox, cock-end, cunt. After an encounter with such a person, we often seek out allies to help talk us down from the disturbance. We can spend hours discussing our neural enemies, listing all the ways they're awful, and it feels disgusting and delicious and is *such* a relief.

We organise much of our lives around reassuring ourselves about the accuracy of the hallucinated model world inside our skulls. We take pleasure in art, media and story that coheres with our models, and we feel irritated and alienated by that which doesn't. We applaud cultural leaders who argue for our rightness and, on encountering their opposite, feel defiled, disturbed, outraged and vengeful, perhaps wishing failure and humiliation on them. We surround ourselves with 'like-minded' people. Much of our most pleasurable social time is spent 'bonding' over the ways we agree we're right, especially on contentious issues. When we meet people who have unusually similar models to us, we can talk to them nonstop. It's so blissful, reassuring ourselves like this, that time itself seems to vanish. We crave their company and put photos of them – arms across shoulders, smiles in beams – on our fridges and social-media feeds. They become friends for life. If the circumstances are right, we fall in love.

It's important to note, of course, that we don't defend all our beliefs like this. If someone approached me and argued that they can prove that every bipartite polyhedral graph with three edges per vertex has a Hamiltonian cycle, or that the Power Rangers could beat the Transformers in a fight, it would have little effect on me. The beliefs we'll fight to defend are the ones which we've formed our identity, values and theory of control around. An attack on these ideas is an attack on the very structure of reality as we experience it. It's these kinds of beliefs, and these kinds of attacks, that drive our greatest stories.

Much of the conflict we see in life and story involves exactly these model-defending behaviours. It involves people with conflicting perceptions of the world who fight to convince each other of their rightness, to make it so their opponent's neural model of the world matches theirs. If these conflicts can be deep and bitter and never-ending, it's partly because of the power of naive realism. Because our hallucination of reality seems self-evident, the only conclusion we can come to is that our antagonist, by claiming to see it differently, is insane, lying or evil. And that's exactly what they think of us.

But it's also by these kinds of conflicts that a protagonist learns and changes. As they struggle through the events of the plot, they'll usually encounter a series of obstacles and breakthroughs. These obstacles and breakthroughs often come in the form of secondary characters, each of whom experiences the world differently to them in ways that are specific and necessary to the story. They'll try to force the protagonist to see the world as they do. By grappling with these characters, the protagonist's neural model will be changed, even if subtly. They'll be led astray by antagonists, who'll represent perhaps darker and more extreme versions of their flaw. Likewise, they'll learn valuable lessons from allies, who are often the embodiment of new ways of being that our hero must adopt.

But before this dramatic journey of change has begun, our protagonist's neural model will probably still be convincing to them, even if it is, perhaps, beginning to creak at its edges – there might be signs that their ability to control the world is failing, which they frantically ignore; there might be portentous problems and conflicts which rise and waft about them. Then something happens . . .

Good stories have a kind of ignition point. It's that wonderful moment in which we find ourselves sitting up in the narrative, suddenly attentive, our emotions switched on, curiosity and tension sparked. This often occurs when we sense an unexpected change has taken place in the plot that sends tremors to the core of the protagonist's flawed theory of control. Because it goes to the heart of their particular flaw, this event will cause them to behave in an unexpected way. They'll overreact or do something otherwise odd. This is our subconscious signal that the fantastic spark between character and event has taken place. The story has begun.

Typically, as their theory of control is increasingly tested and found wanting, the character will lose control over the events of the plot. In an archetypal tale, the more they struggle to regain control, the more trouble and chaos they'll often cause. The drama that is triggered compels the protagonist to make a decision: are they going to fix their flaw or not? *Who are they going to be?*

The cultural model that the butler Stevens had, in *The Remains of the Day*, was nineteenth-century British. It contained core beliefs about the value of dignity and emotional restraint. His model told him that these attributes were the best way to control his environment – that if you behaved with dignity and emotional restraint you would be safe and ultimately rewarded. This theory of control defined him.

And it had been true, in one place and time. But, when we first met Stevens, all that was changing. The power of the British aristocracy that he and his father served, and to which he owed these values, was fading, as was the power of Britain itself. For Stevens, the main practical consequence of these epochal shifts was that his new employer at Darlington Hall, Mr Farraday, was not an English Lord but an American businessman. This was an unexpected change that would challenge the very foundations of who Stevens was. It's a classic ignition point.

As the story starts, Stevens is struggling to meet the challenge of Farraday's not being able to afford the full complement of fourteen staff. Trying to keep the house running with just four people leads him to make 'a series of small errors in the carrying out of my duties' that vex him. But the arrival of his new boss triggers another problem, one that seems to preoccupy Stevens even more: Farraday's 'unfamiliarity with what was and what was not commonly done in England'. Specifically, that his employer enjoys 'conversation of a light-hearted, humorous sort' and has a 'general propensity to talk with me in a bantering tone'.

This bantering makes Stevens profoundly uncomfortable. It's a direct attack on his identity, his beliefs, his theory of control. Bantering isn't what respectable people did. It isn't how you got on. It isn't dignified. It invites not emotional restraint but emotional warmth, and that way lies chaos.

On the one occasion Stevens tries to make a joke, it fails humiliatingly. He proves reluctant to change his core beliefs and his brain, as brains do, provides him with powerful excuses not to.

> It is quite possible that my employer fully expects me to respond to his bantering in a like manner, and considers my failure to do so a form of negligence. This is, as I say, a matter which has given me much concern. But I must say this business of bantering is not a duty I feel I can ever discharge with enthusiasm. It is all very well, in these changing times, to adapt to one's work to take in duties not traditionally within one's realm; but bantering is of another dimension altogether. For one thing, how would one know for sure that at any given moment a response of the bantering sort is truly what is expected? One need hardly dwell on the catastrophic possibility of uttering a bantering remark only to discover it wholly inappropriate.

2.6

We're all fictional characters. We're the partial, biased, stubborn creations of our own minds. To help us feel in control of the outside world, our brains lull us into believing things that aren't true. Among the most powerful of these beliefs are the ones that serve to bolster our sense of our moral superiority. Our brains are hero-makers that emit seductive lies. They want to make us feel like the plucky, brave protagonist in the story of our own lives.

In order to make us feel heroic, the brain craftily re-scripts our pasts. What we actually 'choose' to remember, and in what form, warps and changes in ways that suit the heroic story it wants to tell. When, in the laboratory, participants split money with anonymous people in ways that they themselves considered unfair, they were found to consistently misre-member their own selfish behaviour, even when offered a financial incentive to recall the truth. 'When people perceive their own actions as selfish,' the researchers concluded, 'they can remember having acted more equitably, thus minimising guilt and preserving their self image.'

Our sense of who we are depends, in significant part, on our memories. And yet they're not to be trusted. 'What is selected as a personal memory,' writes Professor of psychology and neuroscience Giuliana Mazzoni, 'needs to fit the current idea that we have of ourselves.' This isn't simply a matter of strategic forgetting. We rewrite and even invent our own pasts. Work by Mazzoni and others has shown that memories can be detailed, vivid and emotional and yet entirely invented. 'We often make up memories of events that never happened,' she writes. Memories are 'very malleable, they can be distorted and changed easily, as many studies in our lab have shown'.

For the psychologists Professors Carol Tavris and Elliot

Aronson, the most important memory distortions 'by far' are the ones that serve to 'justify and explain our own lives'. We spend years 'telling our story, shaping it into a life narrative that is complete with heroes and villains, an account of how we came to be the way we are'. By this process, memory becomes, 'a major source of self-justification, one the story-teller relies on to excuse mistakes and failings'.

But the hero-maker lie goes far beyond memory. The psychologist Professor Nicholas Epley catches it in action when he asks his business students whether they're inspired to pursue careers in industry for heroic 'intrinsic' reasons – doing something worthwhile, pride in achievement, the joy of learning – or more suspect 'extrinsic' ones – pay, security and fringe benefits – and then say the same for their contemporaries. They give matching results every year. They show, writes Epley, 'a subtle dehumanisation of their classmates. My students think all of these incentives are important, of course, but they judge that the intrinsic motivators are significantly more important to them than they are to their fellow students. "I care about doing something worthwhile," their results say, "but others are mainly in it for the money."'

The hero-maker begins with our automatic and mostly subconscious emotional hunches. Say we have models of the world that include racist or sexist beliefs – that give us subtle sensations of 'no' when we encounter black people or white people or women or men. Because we start out convinced we're a good person, then it only logically follows there *must* be a good reason for our negative feelings. So the hero-maker goes on a mission to find them. And it does a good job. It's convincing. After all, who better to fool us – to know *exactly* what to say to beguile us into believing our most incendiary and partisan instincts are morally justified – than our own mind? If we're a good person, the money we stole from our boss must be because they've been exploiting us. If we're a caring person, our political efforts to degrade the NHS must be an altruistic desire to increase efficiency or patient choice. At least that's my take on that situation. That's the moral truth that feels as inarguably real to me as rocks and trees and double-decker buses, because it's made out of the same stuff as those things. I'm blind to any other reasonable argument – I can't perceive them – because they're not part of my perception.

Everyone who's psychologically normal thinks they're the hero. Moral superiority is thought to be a 'uniquely strong and prevalent form of positive illusion'. Maintaining a 'positive moral self-image' doesn't only offer psychological and social benefits, it's actually been found to improve our physical health. Even murderers and domestic abusers tend to consider themselves morally justified, often the victims of intolerable provocation. When researchers tested prisoners on their hero-maker biases, they found them to be largely intact. The inmates considered themselves above average on a range of pro-social characteristics, including kindness and morality. The exception was law-abidingness. There, sitting in prison, serving sentences precisely because they'd made serious contraventions of the law, they were only willing to concede that, on law-abidingness, they scored about average.

The hero-maker delusion is implicated in more misery, fury

and death than is possible to calculate. Mao and Stalin and Pol Pot believed they were right, as did Hitler, whose last words before shooting himself were, 'The world will be eternally grateful to National Socialism that I have extinguished the Jews in Germany and Central Europe.' Indeed, the brains of even the lowliest Nazis automatically generated reasons why what they were doing was morally correct. In the Holocaust's early stages, ordinary middle-aged Germans were recruited to efforts to exterminate Jews. One, a 35-year-old metal worker, remembered, 'it so happened that the mothers led the children by the hand. My neighbour then shot the mother and I shot the child that belonged to her, because I reasoned with myself that, after all, without its mother the child could not live any longer. It was supposed to be, so to speak, soothing to my conscience to release children unable to live without their mothers.'

Researchers have found that violence and cruelty has four general causes: greed and ambition; sadism; high self-esteem and moral idealism. Popular belief and clichéd stories tend to have it that greed and sadism are dominant. In fact, they're vanishingly small. It's actually high self-esteem and moral idealism – convictions of personal and moral superiority – that drive most acts of evil.

In Gillian Flynn's *Gone Girl*, the antagonist Amy Elliott Dunne is motivated, in part, by her pathologically high self-esteem. She's driven to frame her husband for her murder not because of his affair, precisely, but because of what his affair would do to her perceived reputation. On discovering his infidelity, she writes in her diary,

I could hear the tale, how everyone would love telling it: how Amazing Amy, the girl who never did wrong, let herself be dragged, penniless, to the middle of the country, where her husband threw her over for a younger woman. How predictable, how perfectly average, how amusing. And her husband? He ended up happier than ever. No. I couldn't allow that . . . I changed my name for that piece of shit. Historical records have been altered – Amy Elliott to Amy Dunne – like it's nothing. No, he does not get to win. So I began to think of a different story, a better story, one that would destroy Nick for doing this to me. A story that would restore my perfection. It would make me the hero, flawless and adored. Because everyone loves the Dead Girl.

A hero-maker narrative based on moral superiority is convincingly captured in Graham Greene's *The Power and the Glory*, which is set in Mexico during the persecution of the Catholic Church. When a murderous police lieutenant examines a photograph of a wanted priest, the emotion comes first: 'Something you could almost have called horror moved him'. Next comes the self-justifying memory, followed instantly by a hero-maker narrative that ties it all together so that the killer is reassured he's a moral actor:

he remembered the smell of the incense in the churches of his boyhood, the candles and the laciness and the self-esteem, the immense demands made from the altar steps by men who didn't know the meaning of sacrifice. The old peasants knelt there before the holy images with their arms held out in an attitude of the cross: tired by the long day's labour . . . and the priest came round with the collecting-bag taking their centavos, abusing them for all their small comforting sins, and sacrificing nothing at all in return . . . He said, 'We will catch him.'

A character's conviction in their rightness and superiority is precisely what gives them their terrible power. Great drama often forms itself around a clash of competing hero-maker narratives, one belonging to the protagonist, the other to their foe. Their respective moral perceptions of reality feel utterly genuine to their owners and yet are catastrophically opposed. These are neural worlds that become locked in a fight to the death.

2.7

As irrational as we can be, it's important not to infer from all this that we're incapable of ever thinking straight. Of course, reason has power, people can think sensibly and minds can change. It's relatively rare, though, for people to shift significantly on the beliefs around which they form their identity, such as Ishiguro's butler Stevens's convictions about the value of emotional restraint. It's these brave souls we mythologise in story.

One such real-life hero is the former 'eco-terrorist' Mark Lynas. He belonged to a 'radical cell' of the anarchist environmental group Earth First and would hack down experimental genetically modified crops in the night. Earth First told a kind of David and Goliath story about the world, in which the overwhelming forces of industrialism were bringing about, 'environmental apocalypse. Big corporations and capitalism in general were destroying the earth.' Mark's struggle was against the monstrous machines of profit. 'We were protectors of the land and the inheritors of the natural forces,' he said. 'We were the pixies.'

But when he discovered that the science of genetically modified food didn't confirm what his neural models had been telling him, he went through a painful public conversion. As he did, his brain scripted a new story of the world, one in which he could still feel heroic. He'd once perceived the green movement as the brave, scrappy underdogs. But the more he looked now, the more little David took the form of Goliath. 'Just take the numbers,' he said. 'Greenpeace, the whole international group, is a $150m outfit. Bigger than the World Trade Organisation, and much more influential in terms of determining how people think. And there's very deep networks of money and power and influence there too.'

This division of the world into opposing forces of plucky David and almighty Goliath seems a signature manoeuvre of the hero-making brain. The broad narrative it tells of the world is that we're moral actors, struggling against great, Goliathine odds for the good of our lives and perhaps the world. This is a story that gives our lives meaning. It pulls our eyes from the terrible void above and forces them into the urgent now.

The protagonist of *Citizen Kane* expresses just such a heroic narrative when he's challenged by an antagonist. Although the film begins with the death of Charles Foster Kane, the ignition point for his drama is his inheritance of the family fortune. Kane's models of the world are broken in such a way that he has a desperate craving for approval and attention. It's these specific flaws that ignite his story, when he makes the surprising decision to focus on a failing newspaper his estate acquired in a foreclosure proceeding. On his arrival at the paper, his flawed models, now unleashed, begin to exert their influence. At first, it seems as if they're not flawed at all – quite the opposite. He might be happy to be cavalier with the truth in pursuit of his mission ('You provide the prose poems, I'll provide the war!') but he's campaigning on behalf of the disadvantaged citizenry who, he argues, are being exploited by the captains of capitalism.

But then his wealthy, pro-capitalist former guardian – the aptly named Thatcher – confronts him, outraged at what he perceives as his newspaper's 'senseless attack on everything and everybody who's got more than ten cents in his pocket'. When Thatcher reminds him he's a major stockholder in one of the companies he's been attacking, Kane's hero-maker narrative rears up: 'I am the publisher of the *Inquirer*!' he says. 'As such it's my duty – I'll let you in on a little secret, it is also my pleasure – to see to it that decent, hard-working people in this community are not robbed blind by a group of money-mad pirates because they haven't anybody to look after their interests.'

2.8

A man's new boss likes to joke with him and he doesn't like it. It hardly seems like the stuff of great fiction. But it's of critical importance to the man to whom it happens. It shakes the foundations of the butler Stevens's beliefs about how the world correctly operates and who he should be in it. The model of reality he inhabits, inside his skull, comes under threat. When this unexpected change occurs, he tries to regain control over his external environment. He attempts a joke. In order to tackle the staffing problems his boss has created, he embarks on a road trip to Cornwall in the hope of persuading a talented former housekeeper, Miss Kenton, to rejoin his team.

We soon learn that Kenton possesses the warmth Stevens lacked, and yet another loss caused by his devotion to the ideal of emotional restraint was a potential romance with her. Much of the surface drama in *The Remains of the Day* is organised around Stevens's road trip and our changing perceptions of his relationship with Kenton. But, in its depths, this isn't what the story's really about. Beneath the surface causes and effects of the plot, a deeper parallel process is going on. Stevens is changing. His model of the world is slowly and painfully breaking apart.

It's easy to think that a story's surface events – its twists, chases, explosions – are its point. Because we're experiencing it through the eyes of the characters, we, like them, can become distracted by the drama of these thrilling changeful episodes. But none of them mean anything without a specific person for them to happen to. A shark tank has no meaning without a 007 to fall into it. Even crowd-pleasing tales such as James Bond's rely on character for their drama. Those stories are gripping, not because of the bullets or high-speed ski chases in isolation, but because we want to know how *this* specific person, with *this* specific history and *these* strengths and *these* flaws will get out of it. They'll usually only do so by stretching who they are, by trying something new, by making a some unprecedented effort – by changing. Similarly, a police-procedural drama can feel like a straightforward information-gap heavy mystery about a corpse, but its story usually revolves around questions concerning the motives of various suspects: the always fascinating whys of human behaviour.

Of course, different kinds of story have different levels of emphasis and psychological complexity, but plot without character is just so much light and sound. Meaning is created by just the right change-event happening to just the right person at just the right moment. An opulent ball at the splendid home of the Marquis d'Andervilliers would be of only passing interest if it wasn't happening to the middle-class, status-obsessed and chronically unfulfilled Madame Bovary, who marvels at the wealthy guests' complexions that are the kind that 'comes with money' and 'looks well against the whiteness of porcelain' and which are 'best preserved by a moderate diet of exquisite foodstuffs', while she notices, grimly, that her dreary husband's trousers are 'too tight at the waist'. The ball has meaning only in its effects on Madame Bovary. No matter how bedazzling the events of a plot might be, all story is ultimately about character.

A character's struggle, as we've discovered it so far, has been between themselves and the external world. They inhabit a model of the world, inside their skulls, that they experience as reality. Because that model is flawed, their ability to control the real, external world is harmed. When chaos strikes, their model will begin to break down. They'll slowly lose control and this will bring them into further dramatic conflict with the people and events around them.

But all this is complicated by the fact that characters in story aren't only at war with the outside world. They're also at war with themselves. A protagonist is engaged in a battle fought largely in the strange cellars of their own subconscious mind. At stake is the answer to the fundamental question that drives all drama: who am I?

GRIFFITH COLLEGE DUBLIN
SOUTH CIRCULAR ROAD DUBLIN 8.
Tel: 01 4150490 Fax: (01) 4549265
library@griffith.ie

CHAPTER THREE:

THE DRAMATIC QUESTION

3.0

Charles Foster Kane was a man of the people. He might have inherited a fortune, but he'd decided to reject the life of the mercenary rich. Instead, he chose to be an ally of the downtrodden, even as it went against his own financial interests. As editor of *The New York Daily Inquirer*, he fought for their rights relentlessly. In a bid to serve them even better, he ran for Governor of New York. Who could criticise such a selfless and noble man?

As it turns out, his oldest friend could. In the immediate aftermath of Kane's political campaign we find him alone and sorrowful, pacing his campaign office which is still hectic with streamers and posters and emptiness. He has lost. And then in staggers his best pal Jedediah Leland who, it soon becomes apparent, has been out with his sorrows for a few too many drinks. When Kane ruefully acknowledges 'the people have made their choice', Leland cuts him off. 'You talk about the people as if you owned them, as though they belonged to you,' he says, slurring slightly. 'Goodness. As long as I can remember you've talked about giving the people their rights, as if you could make them a present of liberty. As a reward for services rendered. Remember the working man? You used to write an awful lot about the working man. But he's turning to something called organised labour. You're not going to like that one little bit when you find out it means that your working man expects something as his right, not as your gift. When your precious underprivileged really get together ... I don't know what you'll do. Sail away to a desert island, probably, and lord it over the monkeys.' Kane tells him he's drunk. 'Drunk?' Leland replies. 'What do you care? You don't care about anything except you. You just want to persuade people that you love them so much that they ought to love you back.'

Who was Charles Foster Kane really? That was the challenge that editor Rawlston made to his staff of storytellers at the beginning of *Citizen Kane*. Was he the man his old friend perceived: self-interested, delusional, desperate for approval and attention? Or was he the person his own hero-making brain told him he was: brave, generous and selfless?

Who is this person? This is the question all stories ask. It emerges first at the ignition point. When the unexpected change strikes, the protagonist overreacts or behaves in an otherwise unexpected way, we sit up, suddenly attentive. *Who is this person who behaves like this?* The question then re-emerges every time the protagonist is challenged by another or compelled to make a choice.

Everywhere in the narrative that the question is present, the reader or viewer will likely be engaged. Where the question is absent, and the events of drama move out of its narrative beam, they risk becoming detached – perhaps even bored. If there's a single secret to storytelling then I believe it's this. *Who is this person?* Or, from the perspective of the character, *Who am I?* It's the definition of drama. It is its electricity, its heartbeat, its fire.

Harnessing the energy of the dramatic question means understanding that the answer is not easily found. This is because, even at the best of times, most of us don't actually know who we are. If you were to ask Kane who he was, he'd surely say he was noble and selfless, the opposite of his old friend's drunken accusations. He'd mean it too. But, as the screenplay carefully shows, he'd be wrong.

If Kane was to argue he was noble and selfless, it would be because he'd been listening to a voice in his head – one that was telling him all the ways he was morally right. It's not only psychotics like Mr B who hear such voices. We all do. You can hear yours now. It's reading this book to you, commenting here and there as it goes. Flawed characters, in life and story, are often badly led astray by this inner voice, which is generated by word and speech-making circuitry that is mostly located in the brain's left hemisphere. This voice is not to be trusted.

This isn't simply because it's relaying all those flattering hero-making half-truths to us. The narrator can't be trusted because it has no direct access to the truth of who we really are. It *feels* as if that voice is the thing that's in control of us. It feels as if that voice *is* us. But it's not. 'We' are our neural models. Our narrator is just observing what's happening in the controlled hallucination in our skulls – including our own behaviour – and explaining it. It's tying all the events together into a coherent tale that tells us who we are, why we're doing what we're doing and feeling what we're feeling. It's helping us feel in control of our thrilling neural show. And it's not lying, exactly. It's confabulating. As the philosopher of psychology Professor Lisa Bortolotti explains, when we confabulate 'we tell a story that is fictional, while believing that it is a true story.' And we're confabulating all the time.

This disturbing fact was exposed in a series of famous experiments by neuroscientists Professors Roger Sperry and Michael Gazzaniga. Their studies answered a strange question – what would happen if you planted an instruction into a brain and somehow hid it from the narrator? Say, for example, you managed to insert the instruction WALK into a person's mind. And that person started walking. Without the narrator telling the brain's owner *why* they were walking, how would they explain what they were doing? Would they be like a zombie? Would they just shrug? Or what?

Because most of the circuitry that the narrator relies upon is in the brain's left hemisphere, they'd need to find a way of getting information into the right side and keeping it there, hidden away from it. This would mean recruiting so called 'split-brain' patients – epileptics who, as part of their treatment, had had the wiring that connected their hemispheres cut, but who lived otherwise normal lives.

So that's what they did. They showed a card saying WALK to a split-brain patient such that only their left eye saw it. Because of the way the brain's wired up, this information was sent into the right hemisphere. And, because the wiring between their hemispheres had been cut, that's where it stayed, hidden away from the narrator.

So what happened? The patient stood up and walked. When the experimenters asked him why, he said, 'I'm going to get a Coke.' His brain observed what was happening, in his neural realm, and made up a cause-and-effect story to explain it. It confabulated. It had no idea why he'd really stood up. But it instantly invented a perfectly credible tale to account for the behaviour – a tale that its owner unquestioningly believed.

This happened again and again. When a woman's silent hemisphere was shown a picture of a pin-up girl she giggled. She blamed it on their 'funny machine'. When another woman's silent hemisphere was shown a video of a man being pushed into a fire, she said, 'I don't really know why, but I'm kind of scared. I feel jumpy. I think maybe I don't like this room. Or maybe it's you. You're making me nervous. I know I like Dr Gazzaniga, but, right now, I'm scared of him.'

The job of the narrator, writes Gazzaniga, is to 'seek explanations or causes for events'. It is, in other words, a storyteller. And facts, while nice, don't really matter to it: 'The first makes-sense explanation will do.' Our narrator has no wired-in access to the neural structures that that are largely (or wholly, depending on who you ask) controlling how we feel and what we do. Because the narrator exists separately from the circuits that are the true causes of our emotions and behaviour, it's forced to rapidly hash together any makes-sense (and usually heroic) story it can about what we're up to and why.

It's because of such findings, writes Professor Nicholas Epley, that 'no psychologist asks people to explain the causes of their own thoughts and behaviour anymore unless they're interested in storytelling'. It's why a neuroscientist colleague of Professor Leonard Mlodinow said that years of psychotherapy had allowed him to construct a helpful story about his feelings, motivations and behaviour, 'but is it true?

Probably not. The real truth lies in structures like my thalamus and hypothalamus, and my amygdala, and I have no conscious access to those no matter how much I introspect.'

The terrible and fascinating truth about the human condition is that none of us really know the answer to the dramatic question as it pertains to ourselves. We don't know why we do what we do, or feel what we feel. We confabulate when theorising as to why we're depressed, we confabulate when justifying our moral convictions and we confabulate when explaining why a piece of music moves us. Our sense of self is organised by an unreliable narrator. We're led to believe we're in complete control of ourselves, but we're not. We're led to believe we really know who we are, but we don't.

This is why life can be such a vexing struggle. It's why we disappoint ourselves with behaviour that's mysterious and self-destructive. It's why we shock ourselves by saying the unexpected. It's why we find ourselves telling ourselves off, giving ourselves pep talks or asking, 'What the hell was I thinking?' It's why we despair of ourselves, wondering if we'll ever learn.

In stories, the dramatic question has the power to unfold so unexpectedly and endlessly because the protagonists themselves don't know the answer. They're discovering who they are, moment by moment, as the pressure of the drama is applied. And, as the plot turns, they're often surprised by who they turn out to be. Every time you read something like 'she heard herself say' or 'he found himself doing', these forces are likely at play. Characters – and readers and viewers – are being shown fascinating new answers to the dramatic question.

Often, characters are such a mystery to themselves that they seem in complete ignorance of the truth of their own feelings and motives. In *The Idea of Perfection*, Kate Grenville brilliantly exposes the gap between a character's confabulation and the reality of who they are in an encounter between married Felicity Porcelline and her local butcher, Alfred Chang. Felicity is convinced that Alfred's in love with her. She feels so awkward about the situation that she's taken to dawdling outside his shop until another customer arrives that she can enter with. One evening, when Felicity turns up after hours to ask a favour, she finds herself alone with him. The scene that unfolds causes us to doubt Felicity's confabulation of who, exactly, desires who.

When Felicity first spots Alfred she feels a 'a little pulse of something . . . like apprehension, or stage-fright, but it was not those'. Her narrator provides an immediate confabulation to explain this acute sensation: 'It was knowing he was in love with her.' Felicity's eyes prowl Chang's face and body, noticing an opening in his shirt. 'She could actually see a crease of honey-coloured stomach and his neat little navel.' As they talk, she finds herself calling him by his first name. 'She had never done it before and she did not know why she had done so now. It would only encourage him.' When he hoists his trousers up, she sees 'a bulge just there. They were frayed just there, too, around the zip. She looked away, naturally, but could not help noticing. It was really very badly frayed. She heard herself giggle.' She makes herself 'smile slightly, the way she knew smoothed out the skin of her face in a nice way'. Commenting on his family photos, she surprises herself.

'They're lovely photos, she heard herself gushing. So . . . intimate. That was not really the word she had meant. Intimate. It did not sound quite right. She hurried on, before the word could become large in the silence.'

At this stage, it would be an unbelievable shock to Felicity to learn that she ends up in bed with Alfred. But it wouldn't surprise you or me. That 'little pulse of something' she felt on spotting him was her own lust. Like Jedediah Leland in his coruscating view of his old friend Kane, we can clearly see answers to the dramatic question to which Felicity herself is blind. The scene works so brilliantly because the answer keeps changing, paragraph by paragraph, line by gripping line.

3.1

For years I've struggled with cravings and addictions. In middle age, I battle with food. Because the culture I'm immersed in is obsessed with bodily perfection and youth, and because that culture is in me, I find myself engaged in a hopeless quest to make my stomach appear as it did when I was eighteen. What I've discovered, as I've waged these tedious wars against myself, is that who I am seems to be in constant flux.

On the Monday morning following a large roast dinner, I am Captain Abstemious himself, determined, rigid, positively Victorian in my values. I will clean out my cupboards and sort out my life. But by 17:00 Wednesday evening, Captain Abstemious has vanished. In his place stands Billy Pillock Jnr who believes it's pathetic for a man in his forties to worry about a bit of belly flub. He's earned a bit of a treat, with the week he's been having. And what sort of person are you anyway, beating yourself up over a mouthful of Roquefort? How joyless, how vain, how positively Victorian! The problem of self-control, I've come to think, isn't really one of willpower. It's about being inhabited by many different people who have different goals and values, including one who's determined to be healthy, and one who's determined to be happy.

As well as having models of everything in the world, inside our heads, we have different models of self that are constantly fighting for control over who we are. At different times, under different circumstances, a different version of us becomes dominant. When it does, it takes over the role of neural narrator, arguing its case passionately and convincingly and usually winning. Beneath the level of consciousness we're a riotous democracy of mini-selves which, writes the neuroscientist Professor David Eagleman, are 'locked in chronic battle' for dominion. Our behaviour is 'simply the end result of the battles'. All the while our confabulating narrator 'works around the clock to stitch together a pattern of logic to our daily lives: what just happened and what was my role in it?' Fabrication of stories, he adds, 'is one of the key businesses in which our brain engages. Brains do this with the single minded goal of getting the multifaceted actions of the democracy to make sense.'

The truth of our multiplicity is revealed in a condition known as Alien Hand Syndrome. In these patients a behaviour that would usually have been suppressed takes independent control of a limb. The German neurologist Dr Kurt Goldstein recalled a woman whose left hand 'grabbed her own neck and tried to throttle her, and could only be pulled off by force'. The American neurologist Dr Todd Feinberg saw a patient whose hand 'answers the phone and refuses to surrender the receiver to the other hand'. The BBC told of a patient whose doctor asked why she was undressing. 'Until he said that, I had no idea that my left hand was opening up the buttons of my shirt,' she said. 'So I start rebuttoning with the right hand and, as soon as I stopped, the left hand started unbuttoning them.' Her alien hand would remove items from her handbag without her knowing. 'I lost a lot of things before I realised what was going on.' Professor Michael Gazzaniga describes a patient who 'grabbed his wife with his left hand and shook her violently, while with the right hand trying to come to his wife's aid.' One day Gazzaniga saw that patient's left hand pick up an axe. 'I discreetly left the scene.'

Our multiplicity is revealed whenever we become emotional. When we're angry, we're like a different person with different values and goals in a different reality than when we feel nostalgic, depressed or excited. As adults, we're used to such weird shifts in selfhood and learn to experience them as natural and fluid and organised. But for children, the experience of transforming from one person to another, without any sense of personal volition, can be deeply disturbing. It's as if a wicked witch has cast an evil spell, magicking us from princess to witch.

In his pioneering classic *The Uses of Enchantment* the psychoanalyst Professor Bruno Bettelheim argues that making sense of such terrifying transformations is a core function of fairytales. A child can't consciously accept that an overwhelming mood of anger may make him 'wish to destroy those on whom he depends for his existence. To understand this would mean he must accept the fact that his own emotions may so overpower him that he does not have control over them – a very scary thought.'

Fairytales take those scary inner selves and turn them into fictional characters. Once they've been defined and externalised, like this, they become manageable. The story these characters appear in teaches the child that, if they fight with sufficient courage, they can control the evil selves within them and help the good to become dominant. 'When all the child's wishful thinking gets embodied in a good fairy; all his destructive wishes in an evil witch; all his fears in a voracious wolf; all the demands of his conscience in a wise man encountered on an adventure; all his jealous anger in some animal that pecks out the eyes of arch-rivals – then the child can finally begin to sort out his contradictory tendencies,' writes Bettelheim. 'Once this starts, the child will be less and less engulfed by unmanageable chaos.'

Of course, the idea of multiplicity has limits. We don't transform completely, like Jekyll and Hyde. We have a core personality, mediated by culture and early life experience, which is relatively stable. But that core is a pole around which we're constantly, elastically moving. How we behave, in any given moment, is a combination of personality and situation.

In well-told stories, characters reflect this. They're 'three-dimensional' or more. They're both recognisably who they are and yet constantly shifting as their circumstances change. A scene in John Fante's *Ask the Dust* captures this well. The novel tells of young Arturo Bandini's unrequited love for waitress Camilla Lopez. In one dark and dynamic sequence, the character of Bandini comes alive in all his convincing multiplicity when he visits Camilla at the Columbia Buffet, where she works.

Watching her laughing with some male customers Bandini bristles with jealousy. He politely beckons her over, telling himself, 'Be nice to her, Arturo. Fake it.' He asks to see her later. She tells him she's busy. He 'gently' requests she postpone her engagement. 'It's very important that I see you.' When she declines again, his angry self rears up. He pushes his chair back and shouts, 'You'll see me! You little insolent beerhall twirp! You'll see me!' He stalks out and waits by her car, telling himself 'she wasn't so good that she could excuse herself from a date with Arturo Bandini. Because, by God, I hated her guts.'

When she finally emerges, Bandini tries to force her to leave with him. After a tussle she escapes with a barman. Bandini is left in a stew of self-hatred.

Bandini, the idiot, the dog, the skunk, the fool. But I couldn't help it. I looked at the car certificate and found her address. It was a place near 24th and Alameda. I couldn't help it. I walked to Hill Street and got aboard an Alameda trolley. This interested me. A new side to my character, the bestial, the darkness, the unplumbed depth of a new Bandini. But after a few blocks the mood evaporated. I got off the car near the freight yards. Bunker Hill was two miles away, but I walked back. When I got home I said I was through with Camilla Lopez forever.

In this passage, Fante shows Bandini in all his contradiction and multiplicity. One moment he loves her, the next he hates her. One moment he's swollen with arrogance, the next he's a skunk and a fool. His decision to stalk her is an urge that plumes out of his subconscious. When it suddenly dissipates, he doesn't question the madness of his own sudden reversal.

This is a man riding the tumultuous forces of his own hidden brain. He's only barely managing to keep his illusion of self-control intact. It's hard to read this scene without recalling those alien hands discharging unrepressed wills, unbuttoning, throttling and grabbing for the axe. It's structurally effective because of its adherence to cause and effect, with one event leading to another unexpected event which leads to another, and so on. It's meaningfully effective because it keeps asking and answering that essential dramatic question: who is Bandini?

3.2

Nobody can agree which tree is the most photographed in the world. Some say it's a Cypress in Monterey, California, others a Jeffry Pine in nearby Yosemite and others still a Willow in Lake Wanaka, New Zealand. Even if you've never seen them, you can probably guess what these trees look like. They stand alone in endless vistas of water, sky or rock.

Millions of brains have been attracted to the hidden and half-hidden truths that emit from these solitary trees. They triggered something in the photographers' subconscious which responded by giving their owners a pleasurable hit of *feeling.* Lonely, brave, relentless and beautiful, those who stop and snap are not taking pictures of trees, but of themselves.

What these photographs reveal is that human consciousness works on two levels. There's the top level on which occurs the drama of our day-to-day lives – that meeting of sight, sound, touch, taste and smell which is narrated by the hero-making inner voice. And then, beneath that, there's the subconscious level of the neural models, a stewing night ocean of feelings, urges and broken memories in which competing urges engage in a constant struggle for control.

The stories we tell also work on these levels. They operate 'in two realms', writes the psychologist Professor Jerome Bruner, 'one a landscape of action in the world', the other a landscape of the mind in which the 'protagonists' thoughts and feelings and secrets play themselves out'. On the plot's conscious top layer we experience the visible causes and effects of the drama. Then there's the story's subconscious that heaves beneath the visible. It's a place of symbolism and division, in which characters are multiple and contradictory and surprising, even to themselves.

Some of the most moving moments in story come when the second subconscious layer erupts into the first. Jill Soloway's TV drama *Transparent* brought me to tears when the character Josh Pfefferman suddenly revealed himself in a way that surprised even him. The series tracks the ramifications of a family patriarch's decision to transition to a woman, from Mort to Maura. Josh, Maura's son, is jovial, wry and essentially decent. He's a record company executive and thoroughly modern, always wanting to be supportive of Maura's journey.

But things start slipping for Josh. Towards the end of the second series, he's driving with some band members and starts uncharacteristically ranting: 'Look at this traffic! They time it out so you can't get anywhere. It's a fucking conspiracy.' He honks his horn at other drivers. 'Fucking go, you piece of shit! They're fucking boxing me in!' He's losing control. The woman beside him insists he pulls over. Josh is hyperventilating.

Sometime afterwards, he calls to see his mother Shelley only to discover she's out. Her new boyfriend Buzz lets him in. 'Nothing's adding up,' Josh confides to Buzz. 'I thought stuff would add up by now, but everything's slipping through.' Buzz, with his grey ponytail and hippy shirt, is of a different generation to Josh. His model of the world comes from an earlier time. He suggests Josh is in 'shock' about the 'loss' of his father. Josh pushes back. Buzz doesn't get it, nobody has died. 'You think I miss Mort?' he asks, irritated.

'What do you think?' says Buzz.

'Well, it's like politically incorrect to say that you miss someone who has transitioned, so . . .'

'This isn't about correct, Joshua, this is . . . This is about grieving. Mourning. Have you grieved and mourned the loss of your father?'

'Him? Like losing him? No, I'm . . . I don't know how to do that.'

There's a moment of silence. Josh crumbles into the arms of the older man and sobs.

In well-told stories, there's a constant interplay between the surface world of the drama and the subconscious world of the characters. The bedlam that takes place on the top often has seismic subconscious ramifications for who the character is beneath. As the psychologist Professor Brian Little writes, 'All individuals are essentially scientists erecting and testing their hypothesis about the world and revising them in the light of their experience.' As these subtle revisions in who they are take place, on the subconscious second level, the answer to the dramatic question changes. And as their character changes this, in turn, alters their behaviour on the surface level of the drama. And so and so on.

This is how plots develop as they should – from character. At the ignition point, when the drama starts flying at them, their subconscious model of the world receives its first serious crack. They'll try to reimpose control. These attempts will fail. They might even make the situation worse. With their neural model of the world increasingly foundering, they enter a subconscious state of panic and disorder.

As their models fracture and break down, previously repressed wills, thoughts and versions of self rise up and become dominant. This can be seen as the brain's experiments in novel ways of controlling its environment. They might find themselves behaving in ways they weren't expecting, as Arturo Bandini did when he unexpectedly turned stalker. These unexpected behaviours might cause them to learn something about themselves, as Josh Pfefferman did when he collapsed sobbing.

Some of the most memorable scenes in drama allow us to watch the dramatic question battle itself in the mind of the character. In such scenes, the character appears divided and in a state of internal conflict. What they're saying, for example, might contradict how they're behaving in ways that show they're manifesting as two different versions of self at once. We can't quite tell what they're going to do next. Who they are is changing before our eyes.

And so the plot moves on, in all its depth, truth and unpredictability, each new development coming from character. Inch by inch, scene by scene, characters and plot interact, each altering the other. Throughout the plot, as the character confronts the fact that they're failing to control the world, they're gradually forced to readdress their deepest beliefs about how it works. Their precious theory of control comes under question. Beneath the level of consciousness, they're compelled to repeatedly ask themselves that fundamental dramatic question: who am I? Who do I need to be in order to make this right?

This is the process that drives Robert Bolt and Michael Wilson's cinematic masterpiece *Lawrence of Arabia*. An approximate definition of Lawrence's flaw would be something like *vanity that manifests as rebellion*. He's rather insolent and self-important. This is how he controls the world of people around him. It's how he makes himself feel superior – in one early scene, he showily extinguishes a lit match with his bare fingers. When we meet him he's a lieutenant in the British Army during the First World War. He fails to salute his superior, General Murray, who complains, 'I can't make out if you're bloody bad-mannered or just half-witted.'

'I have the same problem, sir,' replies Lawrence with a supercilious lilt.

'Shut up.'

'Yes sir.'

Lawrence is sent to the Middle East on an intelligence mission. The ignition point comes when he's journeying through the desert to begin his work and his local guide is shot dead by an Arab leader, Sherif Ali, because he drank from his well. This unexpected change connects specifically with Lawrence's flawed theory of control, which is based around rebelliousness and vanity. He reacts in an unexpected way. His flaw causes him not to flee or grovel for his life but to grandiosely berate the killer: 'Sherif Ali! So long as the Arabs fight tribe against tribe, so long will they be a little people, a silly people, greedy, barbarous and cruel – as you are.' Gone is the insolent wally of the previous scenes. The dramatic question has been posed.

After Lawrence experiences a brutal attack on the Arabs by their enemies, the Turks, his rebellious vanity rises again. He becomes engaged in the Arabs' fight and suggests they all trek through the hellish Nefud desert and launch a surprise attack on a Turkish stronghold. On the journey, Lawrence's rebellious vanity kicks up when, against everyone's advice, he insists on making an insanely dangerous journey back into the desert to rescue a lost Arab. When he returns with the man, the Arabs ecstatically cheer him. Once again, the first layer of drama affects the second layer of subconscious. His theory of control – that you got what you wanted with vain rebelliousness – has been proven right. And so he becomes yet more vain and rebellious. He's accepted into the tribe. In a deeply symbolic moment, Sherif Ali, the man who shot his guide, burns his western clothes and dresses him in 'the robes of a Sherif'. When Lawrence leads the Arabs on a successful assault on the Turkish stronghold, his vanity soars even more.

And yet, beneath the level of the surface drama, things have started cracking. Just before the successful assault, Lawrence had been compelled to execute a man in order to prevent factions of his Arab force attacking one another. After the assault, he accidentally leads his men into quicksand. One of them dies. These experiences disturb him. When he finally makes it out of the desert, to the shores of the Suez Canal, a motorcyclist on the opposite bank spots him. Curious about this strange white man in Arab robes emerging from the desert, the motorcyclist shouts across the water: 'Who are you? Who are you?' As the question fills the baking air, the camera freezes on Lawrence's troubled face.

Who is he? Is he the man his flaw of rebellious vanity tells him he is? Is he extraordinary? Or is he just ordinary? This simple question underpins every gripping scene of the film. So far, he's proved to be mostly extraordinary. His theory of control has worked. His vain rebelliousness has led him to success after success. We cheer when he berates the killer Sherif Ali! We applaud when he rescues the fallen soldier! We roar when he wins his battle! But if this was all there was to the story, it wouldn't have won seven Academy Awards.

The pressure of the drama is beginning to crack Lawrence's model of the world. Adherence to his theory of control might be leading him to great victories but it's also causing him deep subconscious distress. Our first real clue about these dark changes that are happening to him arrives when he comes in from the desert and General Murray promotes him and asks him to go back. Lawrence refuses. 'I killed two people,' he explains. 'I mean, two Arabs. One was a boy. That was yesterday. I led him into quicksand. The other was a man . . . I had to execute him with my pistol. There was something about it I didn't like.'

'Well, naturally,' says Murray.

'No, something else,' he says. 'I enjoyed it.'

In this highly dramatic scene, we see Lawrence divided. He's learned to control the world by adherence to a vanity that manifests as rebellion. This theory of control has driven him to huge success. It's enabling him to become an extraordinary man. But it's also led to unexpected effects. He has glimpsed what he's turning into, and what 'success' actually means, and it's terrified him.

But the military chiefs ignore Lawrence's pleas. And they know just how to convince a vain man like him – by shoring up his leaking theory of control. They tell him his feats in the desert were superhuman and recommend him for a medal. He's a brilliant soldier, they say. He's *extraordinary*. Precisely because of the nature of Lawrence's flaw, their manipulations work. He returns to the desert more vain and rebellious than ever. He leads an attack on a Turkish train. The Arabs loot it and hail him almost as a living god: 'Lawrence! Lawrence! Lawrence!'

His flaw deepens. He begins demanding the impossible of his men – 'My friends, who will walk on water with me?' When Sherif Ali protests that he's asking too much of them, Lawrence pushes back: 'Whatever I ask them to do can be done . . . Do you think I'm just anybody, Ali? Do you?'

By now Lawrence has become so vain and rebellious he behaves as if he has magical powers. With a nervous Sherif Ali at his side, he swans into a Turkish garrison, splashing through puddles, utterly convinced he won't be seen despite his glaring whiteness. 'Do you not see how they look at you?' Ali hisses.

'Peace, Ali,' he replies. 'I am invisible.'

But he's not invisible. Lawrence is caught and brutally tortured. His beating is such that he's forced to realise his theory of control was wrong. His most fundamental beliefs about who he was were mistaken, and catastrophically so. Back at base, still bleeding from his wounds, he hands General Murray a written request to leave Arabia.

'For what reason?' demands Murray.

'The truth is,' he says, 'I'm an ordinary man.' But Murray knows how to get around him. 'You're the most extraordinary man I've ever met.'

'Leave me alone,' begs Lawrence. 'Leave me alone.'

'Well that's a feeble thing to say.'

'I know I'm not ordinary.'

'That's not what I'm saying.'

'Alright!' says Lawrence. 'I'm extraordinary. What of it?'

Soon afterwards, in the film's most iconic sequence, Lawrence leads his Arab army in a gruesome attack on fleeing Turks. 'No prisoners!' he yells. 'No prisoners!' When his handgun runs out of bullets, he starts madly slashing at people with his dagger. Sherif Ali, the man he berated at as 'barbarous' and a 'murderer' at the film's start, begs him to stop. Soaked in blood, surrounded by fresh corpses, Lawrence lifts the gory blade of his knife and gazes in horror at his reflection.

Stories such as this are like life itself, a constant conversation between conscious and subconscious, text and subtext, with causes and effects ricocheting between both levels. As incredible and heightened as they often are, they also tell us a truth about the human condition. We believe we're in control of ourselves but we're continually being altered by the world and people around us. The difference is that in life, unlike in story, the dramatic question of who we are never has a final and truly satisfying answer.

3.3

Tragedies such as *Lawrence of Arabia* can be especially useful, for the purposes of analysis, because the causes and effects of character change tend to have greater emphasis in the narrative and are therefore clearer to see. But all archetypal stories are like this, even if the process is less overt in some. They're about flawed selves being offered the opportunity to heal. Whether their endings are happy or otherwise depends on whether or not they take it. If they choose to heal, like Ebenezer Scrooge in Charles Dickens's *A Christmas Carol* or, say, Charlie Simms and Lieutenant Colonel Frank Slade, the twin protagonists of Bo Goldman's Academy Award-winning *Scent of a Woman*, the audience will be profoundly cheered. But whatever happens, we're usually left in little doubt as to what conclusion the writer wanted us to come to. In the closing scenes, the dramatic question will have been answered. We'll leave the story with that lovely emotional sense that something, perhaps just beyond the level of conscious comprehension, has been completed.

Modernist stories are different. Whilst they're built from the same dance between surface drama and subconscious change, their causes and effects are often left ambiguous. Character change occurs, but it's less clear how these changes are being triggered by the drama and what message we're supposed to glean from them. More space is left for the reader to insert their own interpretations into the text.

Franz Kafka's short story 'The Passenger' shows an enigmatic movement of cause and effect between consciousness and subconscious. It tells of a man on a tram feeling uncertain about himself and his place in the world. He becomes lost, for a moment, in the abstract physical details of a woman waiting to disembark – the position of her hands, the shape of her nose, the shadow her ear makes against her skull. These conscious observations trigger something deep in his subconscious. He asks, 'How is it that she is not astonished at herself, that she keeps her mouth closed, and expresses nothing of any wonderment?' In a way that recalls eastern story forms such as Kishōtenketsu, the reader is invited to ponder how one level connects to the other and thereby bring them into harmony.

Virginia Woolf's *Mrs Dalloway* tracks such movements between consciousness and subconscious in longer form, as it follows a day in the life of eponymous Clarissa, and various characters orbiting her, as she prepares for and hosts a party. The story is told not as if the protagonist is talking out loud to the reader, as is common in first-person narratives. Rather, it's as if we're privy to her inner narrator as it bounces between the external and internal – from event in the world to thought, memory, to sudden revealing insight – bringing it all together into a compelling and believable composite of self.

In a similar style, Knut Hamsun's *Hunger* tracks its unnamed protagonist's struggle to survive mentally and physically while trying to earn money as a writer. Published in 1890, it's a stunningly prescient exploration of human cognition. The central character, who ruefully describes himself as 'nothing but a battleground for invisible forces' is thrown relentlessly between the two levels of cause and effect. On seeing an attractive woman he becomes 'possessed by a strange desire' to frighten her and makes 'stupid faces' behind her back: 'No matter how much I told myself I was acting idiotically, it did not help.'

One morning, for some unknowable reason, the noises of the street send his mood soaring. 'I was powerful as a giant and could stop a wagon with my shoulders . . . I started to hum for pure joy and for no particular reason.' In desperation, he tries to pawn a tattered blanket and is humiliated when the pawnbroker sends him away. After taking it back home: 'I acted as though nothing had happened, spread the blanket out again on the bed, smoothed out the wrinkles as I always did, and tried to erase every trace of my last action. I couldn't possibly have been in my right mind when I decided to try this filthy trick. The more I thought of it, the more irrational it seemed. It must have been some failure of energy far inside that had caught me off guard.'

Generations before science caught up, Hamsun showed how we are multiple and confabulatory, skating on the thin ice of sanity, all of us a battleground for the invisible forces of our own subconscious minds.

3.4

It's not uncommon for a character to *want* something on the conscious level and yet subconsciously *need* something entirely different. As the story theorist Robert McKee writes, 'the most memorable, fascinating characters tend to have not only a conscious but an unconscious desire. Although these complex protagonists are unaware of their subconscious need, the audience senses it, perceiving in them an inner contradiction. The conscious and unconscious desires of a multidimensional protagonist contradict each other. What he believes he wants is the antithesis of what he actually but unwittingly needs.'

Alan Ball's Academy Award-winning screenplay *American Beauty* focuses on just such a character. When we meet 42-year-old Lester Burnham, he's bullied by his boss, his daughter and especially his disdainful and unfaithful wife. Miserable and trapped, Lester suffers a midlife crisis, deciding that happiness lies in his becoming young and carefree again. He buys a fast car, starts working out in his garage, finds a job at a drive-through burger restaurant and smokes marijuana. He stands up to his boss and wife. Much of the surface-level plot is taken up with Lester's blackly comic attempts at sleeping with his daughter's best friend, the apparently street-wise and experienced Angela.

When he finally gets the opportunity to do so we're shown the contradiction between his shallow, short-term conscious desires and his deep subconscious needs. Lying half-naked beneath him, Angela confesses she's not as experienced as she'd appeared: 'This is my first time.'

'You're kidding,' says Lester. He crumbles, refusing to carry on. Angela becomes upset. Lester wraps her in a blanket and holds her as she sobs – a responsible adult, finally.

While Lester *wanted* to be young again, what he'd *needed* was to mature and become truly powerful. In this touching and revelatory moment, as a better version of his self bubbles up from his subconscious, we realise that the answer to the dramatic question has suddenly flipped to its opposite.

The scene has additional power because it doesn't only show a transformation in who we understand Lester to be. We see Angela in a new way too. In all great stories, each major character is altered somehow by their interpersonal encounters. As they clash, they send each other spinning outwards, only to clash again in new and altered ways, and then spin out again, and meet again and so on and so on, out across the plot, in an elegant and gripping dance of change.

3.5

Story time is compressed time. An entire life can be told in the space of just ninety minutes and still somehow feel complete. It's this compression that's the secret of arresting dialogue. The words characters speak should both sound true and writhe with meaning, making for a rich source of data for the model-making brain. Speech should be crammed with deep facts that can be greedily absorbed by readers and viewers, whose hyper-social brains rapidly construct models of the fictional characters' minds.

Some of the most famous lines of dialogue in film history derive their power from the fact that they're so dense with narrative information it's as if the entire story is packed into just a few words:

I love the smell of napalm in the morning.
        Apocalypse Now, *Francis Ford Coppola, John Milius, Michael Herr*

I wish I knew how to quit you.
        Brokeback Mountain, *Larry McMurtry and Diana Ossana via Annie Proulx*

I'm as mad as hell, and I'm not going to take this anymore.
        Network, *Paddy Chayefsky*

The greatest trick the devil ever pulled was convincing the world he didn't exist.
        The Usual Suspects, *Christopher McQuarrie*

I'm just a girl, standing in front of a boy, asking him to love her.   Notting Hill, *Richard Curtis*

These go to eleven.
        This is Spinal Tap, *Rob Reiner, Christopher Guest, Michael McKean, Harry Shearer*

I am big! It's the pictures that got small.
     Sunset Boulevard, *Billy Wilder, Charles Brackett, D. M. Marshman Jr*

You're gonna need a bigger boat.   Jaws, *Peter Benchley*

All the principles of storytelling combine into the art of dialogue. Dialogue should be changeful, it should want something, it should drip with personality and point of view, and it should operate on the two story levels – both conscious and subconscious. It can give us clues about everything we need to know about the character: who they are, what they want, where they're going, where they've been, their social background, their personality, their values, their sense of status, the tension between their true self and the false front they're presenting, their relationships to other characters, the secret torments that will drive the narrative forwards.

Take this opening monologue from the TV series *Marion and Geoff* by Rob Brydon and Hugo Blick. How much do we learn in just eighty-three seconds of screen time about the taxi driver Keith Barrett?

KEITH: [*sliding into his car seat*]: *Good morning, good morning! Another day, another dollar.* [*speaks into handheld radio*] *My first pick up please?* [*white noise – he shrugs.*] *I'll just drive around. It's like that some days. You just ease your way into the day.*

[Cut to Keith driving] KEITH: These sleeping policemen are a wonderful idea, but they're a pain in the bloomin' neck, I'll tell you that. I mean, I'm not against them. I would never say that. If they only save one life . . . then probably not very cost-effective.

[Cut] KEITH: It's not that the kids think of Geoff as their father, because they don't. They think of him as an uncle. A special uncle. A new uncle. I like him. If you like someone you like someone, you can't help it. I mean, I actually said to him, 'I don't feel like I've lost a wife, I feel like I've gained a friend.' I would never have met Geoff if Marion hadn't left me. Not a chance of it. We're in different worlds. He's in pharmaceuticals, I'm in cars. Literally – I'm in the car. I bear you no ill, sir. I bear you no ill.

Similarly, how much do we learn in this brief exchange between the ageing salesman Willy Loman and his wife Linda, from Arthur Miller's *Death of a Salesman*?

WILLY: If old man Wagner was alive I'd a been in charge of New York now! That man was a prince, he was a masterful man. But that boy of his, that Howard, he don't appreciate. When I went up north the first time, the Wagner company didn't know where New England was!

LINDA: Why don't you tell those things to Howard, dear?

WILLY (encouraged): I will, I definitely will. Is there any cheese?

3.6

As we move through the plots of our lives, we're not only struggling against unruly, unpredictable and unhelpful versions of self. We're also fighting to manage powerful drives that are wired deeply into us. These are the products of human evolution. Exposing these drives means travelling back tens of thousands of years, to the era in which we became a storytelling animal. The journey's reward is the unburying of ancient yet critical lessons about story, not least the origin and purpose of the dramatic question.

Films and novels are pleasurable – tense, shocking, gut-wrenching, thrilling, suspenseful, satisfying – in large part because of their ancient roots. The emotions we experience, when under the power of story, don't happen by accident. Humans have evolved to respond in certain ways to tales of heroism and villainy because doing so has been critical for our survival. This was especially true back when we were living in hunter-gatherer tribes.

We've spent more than ninety-five per cent of our time on earth existing in such tribes and much of the neural architecture we still carry around today evolved when we were doing so. In this twenty-first century of speed, information and high technology, we still have Stone Age brains. As powerful as culture is, it cannot cancel out or transform these deeply embedded primal forces, but only modulate it. No matter where we come from, East, West, North or South, Pleistocene winds blow in our subconscious minds, touching almost every part of our modern lives, from our codes of morality to the ways we arrange our furniture. One study found people prefer to sleep as far from their bedroom door as possible and with a clear view of it, as if still in a cave and wary of night-time predators. The body's reflexes remain primed for the savannah we once roamed: when someone creeps up and shocks us, the body automatically responds as if being attacked by a prey animal. All over the world, people enjoy open spaces and lawns and prefer trees of a shape, height and canopy similar to that which we evolved amongst. Our Stone Age values also remain strongly evident in stories.

It's testament to the powers of the storytelling brain that many psychologists argue that human language evolved in the first place in order to tell tales about each other. As unlikely as this sounds, it makes sense. Human tribes were big, topping out at around 150 members who'd occupy a large physical territory and live, day-to-day, in clusters of perhaps five to ten families. In order to be functional, it was essential that members of a tribe cooperated – that they shared and helped and worked together, putting the needs of others before their own. But this presented a problem. Humans are *people*. And yet, despite this apparently catastrophic design flaw, ancient tribes excelled at cooperating. Not only did they manage to do so such that they survived for tens of thousands of years, with some still existing today, they're thought to have been far more egalitarian than modern humans. How did they *do* this? How did they control each other's self-interested behaviour so fantastically, without the help a police force, a judiciary or even any written law?

They'd do it with the earliest and most incendiary form of

storytelling. Gossip. People would keep track of everyone else, closely tallying their behaviour. When these gossipy stories concerned a person behaving selflessly – when they put the tribe's needs before their own – listeners would experience a wash of positive emotions and an urge to celebrate them. But when they were told tales of someone being selfish, listeners would experience the emotion of moral outrage. They'd be motivated to act – to punish them, whether by being shamed and mocked, violently attacked or ostracised from the group, which would've been a sentence of death.

This is how stories kept the tribe together as a functional, cooperating unit. They were essential for our survival. And our brains operate in the same way today. We enjoy great books or immersive films because they're activating and exploiting these ancient social emotions. When a character behaves selflessly we experience a deep primal craving to see them recognised by the group as a hero and hailed. When a character behaves selfishly, we feel a monstrous urge to see their punishment. 'Stories arose out of our intense interest in social monitoring,' writes the psychologist Professor Brian Boyd. They work by 'riveting our attention to social information', whether in the form of gossip or screenplay or books, which typically tell of 'heightened versions of the behaviours we naturally monitor'.

Today, like then, the social emotions that are roused by story motivate us to act. Because we can't jump into a cinema screen and throttle the villain ourselves, the urge to act compels us to keep turning the page or watching the screen until our tribal appetites have been satisfied.

Selfless versus selfish is storified as hero versus villain. We're wired to find selfless acts heroic and selfish deeds evil. Selflessness is thought to be the universal basis of all human morality. An analysis of ethnographic accounts of ethics in sixty worldwide groups found they shared these rules: return favours, be courageous, help your group, respect authority, love your family, never steal and be fair, all a variation on 'don't put your own selfish interests before that of the tribe'.

Even pre-verbal babies show approval of selfless behaviour. Researchers showed six- to ten-month-old infants a simple puppet show in which a goodie square selflessly helps a ball up the hill while a baddie triangle tries to force it down. When offered the puppets to play with almost all these children chose the selfless square. Psychologist Professor Paul Bloom writes that 'these were bona fide social judgements on the part of the babies'.

Further evidence of the universality of the selfless–selfish moral axis comes from story. Theorists have also detected these patterns in myth and fiction. The mythologist Joseph Campbell describes the hero's ultimate test as selflessly 'giving yourself to some higher end ... When we quit thinking primarily about ourselves and our own self-preservation, we undergo a truly heroic transformation of consciousness.' Meanwhile, the story theorist Christopher Booker writes that 'the "dark power" in stories represents the power of the ego . . . [and] is immensely powerful and concerned solely with pursuing its own interests at the expense of everyone else in the world.'

These emotional responses exist as neural networks that

can be activated whenever they detect anything, in the environment, that has the rough shape of tribal unfairness. This leaves storytellers free to trigger them in any number of ways. It doesn't have to be a strictly archetypal pattern of selfless hero versus selfish villain. In the opening sequences of *The Grapes of Wrath* we feel outraged not about a human, but a terrible drought that drives the noble, hardworking Joad family out on the perilous road. It's not *fair* that this is happening to them. We root for them as they battle on towards California. We crave the natural justice of their safety.

In *Mrs Dalloway*, Virginia Woolf plays with these instincts delicately. When Clarissa ponders the 'question of love' she has a memory of an old friend, Sally Seton, 'sitting on the floor with her arms around her knees smoking a cigarette' and asks herself 'had not that, after all, been love?' At this point we feel our social emotions jolting. It has the inescapable quality of gossip – this is a *very* interesting new development about Clarissa Dalloway. When we hear their long-ago kiss was 'the most exquisite moment of her whole life . . . The whole world might have turned upside down!' we feel gently outraged that this love was unable to find true expression – it's not fair! We sit up in the narrative. We care.

Less subtle is *Dancer in the Dark*, a screenplay by Lars von Trier that pounds relentlessly on these same tribal instincts. It tells of a poor Czech immigrant, Selma Ježková, who lives with her son in a caravan at the bottom of a policeman's garden. Selma has a degenerative eye condition. She's going blind. She knows that her son, Gene, has the same hereditary condition and if he's not operated on before he turns thirteen, he'll also lose his sight. In order to pay for his operation, Selma saves all the money she can from her dangerous job at a metal-working factory. At great risk to herself, she keeps her failing eyesight a secret. When her disability becomes obvious, and she breaks a machine, she's fired. Luckily, she has almost enough to pay for Gene's operation. But then her policeman landlord, in whom she's confided, steals her money.

Watching *Dancer in the Dark*, I became so engorged with caveman emotion at this raw and inordinate expression of selfish versus selfless, I'd have gladly stepped into the screen and clubbed him to death. That I was desperate to enact his punishment is, once again, no accident. Just as our storytelling brains are wired to valorise pro-social behaviour, we're designed to love watching the anti-social suffer the pain of tribal comeuppance. These darker instincts are also evident in children. Another psychologist's puppet show starred an evil, thieving puppet who was struggling to open a box. A second puppet tried to help the villain whilst a third puppet – the punisher – jumped on the lid, slamming it shut. Even eight-month-olds preferred to play with the punisher. Brain scans reveal that the mere *anticipation* of a selfish person being punished is experienced as pleasurable.

This 'altruistic punishment' of tribal villains is a form of what's known as 'costly signalling'. It's 'costly' because it's difficult to achieve and hard to fake and a 'signal' because its purpose is to influence what other members of the tribe think of them. 'The heroes and heroines of narrative are those who pay the costs of defending the innocent and who punish defec-tors,' writes Professor of English Literature William Flesch.

'Because it is costly, and because bearing those costs is heroic, altruistic punishment is a common characteristic of heroes.' Heroes in archetypal stories are selfless costly signallers. In the face of great personal peril, they kill dragons, blow up Death Stars and rescue Jews from Nazis. They satisfy our moral outrage, and moral outrage is the ancient lifeblood of human storytelling.

In many of our most successful stories, moral outrage is triggered in the early scenes. Watching a selfless character being treated selfishly is a drug of enchantment for the storytelling brain. We almost can't help but care. Selfish versus selfless is also the shape of most human gossip: studies reveal that, not only is gossip universal, with around two-thirds of our conversation being devoted to social topics, most of it concerns moral infractions: people breaking the rules of the group.

All this reveals why the fundamental drive of our films, novels, journalism and plays is the dramatic question. Whether the protagonist we're gripped by is Lawrence of Arabia or a rude dad in some school-gate tittle-tattle, what we ultimately want to know is its answer – who *is* he? The surprising discovery that's been waiting for us, at the destination of our long journey into our evolutionary past, is that all story is gossip.

3.7

Moral outrage isn't the only primal social emotion that's responsible for the pleasure of storytelling. Evolutionary psychologists argue we have two wired-in ambitions: to *get along* with people, so they like us and consider us non-selfish members of the tribe, and also *get ahead* of them, so we're on top. Humans are driven to connect and dominate. These drives, of course, are frequently incompatible. Wanting to get along *and* get ahead of them sounds like a recipe for dishonesty, hypocrisy, betrayal and Machiavellian manoeuvring. It's the conflict at the heart of the human condition and the stories we tell about it.

Getting ahead means gaining status, the craving for which is a human universal. The psychologist Professor Brian Boyd writes, 'Humans naturally pursue status with ferocity: we all relentlessly, if unconsciously, try to raise our own standing by impressing peers, and naturally if unconsciously, evaluate others in terms of their standing.' And we need it. Researchers have found that people's 'subjective well-being, self-esteem, and mental and physical health appear to depend on the level of status they are accorded by others.' In order to manage their status, people 'engage in a wide range of goal-directed activities'. Underneath the noblest plots and pursuits of our lives, in other words, lies our unquenchable thirst for status.

Humans are interested in the status of themselves, and others, to an almost obsessional degree. Studies of gossip in contemporary hunter-gatherer tribes find that, just like the stories that fill the newspapers of great cities and nations, it's dominated by tales of moral infractions by high-status people. Indeed, our preoccupation with the subject stretches back deep into our animal pasts. Even crickets keep a tally of their victories and failures against cricket rivals. Researchers into bird communication have revealed the astonishing fact that not only do ravens listen to the gossip of neighbouring flocks, but they pay especially close attention when it tells of a reversal in another bird's status.

If many animals are similarly status-obsessed, our special interest in it comes partly because human hierarchies are not static but fluid. We have this in common with chimpanzees who, along with bonobos, are our closest cousins. We can infer from this closeness that any habits we share with them probably stretch back to the ancestor we have in common and with whom we split between five and seven million years ago. Chimpanzee alphas have a lifespan at the top of about four to five years. Because status is of existential importance (benefits for chimps and humans include better food, better mating opportunities and safer sleeping sites) and because everyone's status is always in flux, it's a near-constant obsession. This status flux is the very flesh of human drama: it creates running narratives of loyalty and betrayal; ambition and despair; loves won and lost; schemes and intrigues; intimidation, assassination and war.

Chimpanzee politics, like human politics, runs on plots and alliances. Unlike so many other animals, chimpanzees don't only fight and bite their way to the top, they also have to be coalitional. When they reach the heights, they need to adopt a policy of sensitive politicking. Lashing out at those beneath them risks triggering revolt and revolution. 'The tendency of chimps to rally for the underdog creates an inherently unstable hierarchy in which the power at the top is shakier than in any monkey group,' writes the primatologist Professor Frans de Waal. When troop leaders are toppled from their throne, it's usually because a gang of low-status males has conspired against them.

Precisely these patterns of status play haunt human lives and stories. The story theorist Christopher Booker writes of an archetypal narrative form in which low-ranking characters 'below the line' conspire to topple the corrupt and dominating powers above it. 'The point is that the disorder in the upper world cannot be amended without some crucial activity taking place at a lower level,' he writes. 'It is from the lower level that life is regenerated and brought back to the upper world again.' The necessary characteristics to become a human hero mirror those necessary for a chimpanzee to rise to a position of dominance. At the happy ending of an archetypal story, Booker writes, a 'hero and heroine must represent the perfect coming together of four values: strength, order, feeling and understanding.' This same combination of characteristics is required in chimp alphas, whose place on top depends on their balancing straightforward dominance with a will (or at least its appearance) to protect those lower on the ladder.

But if a protagonist learns these four values of heroism at the end of the story, and is therefore rewarded with the ultimate prize of tribal status, that's not how they begin. When we meet them, they're frequently low in the hierarchy – vulnerable, reluctant, trembling in the shadow of Goliath. Just as for our cousins the chimpanzees, our empathy with these underdogs comes naturally. A common feature of our hero-making cognition seems to be that we all tend to feel like this – relatively low in status and yet actually, perhaps secretly, possessing the skills and character of someone deserving of a great deal more. I suspect this is why we so easily identify with underdog heroes at the start of the story – and then cheer when they finally seize their just reward. Because they're *us*.

If this is true, it would also explain the odd fact that, no matter what our level of actual privilege, everyone seems to feel unfairly lacking in status. Biographer Tom Bower writes that Prince Charles is among the chronically dissatisfied, a condition that perhaps isn't helped by his association with billionaires. 'During a recent after-dinner speech at Waddesdon Manor, Lord Rothschild's Buckinghamshire home, Charles complained that his host employed more gardeners than himself; fifteen against his nine.' No matter who we really are, to the hero-making brain we're always poor Oliver Twist: virtuous and hungry, unfairly deprived of status, our bowls bravely offered out: 'Please, sir, I want some more.'

As much as we might feel like the beloved Oliver Twist, we're also wired to despise the cruel higher-status Mr Bumbles that surround us. Even when they're not actually deserving of our wrath, as Dickens's pompous workhouse boss surely is, we naturally dislike them. When people in brain scanners read of another's wealth, popularity, good looks and qualifications, regions involved in the perception of pain became activated. When they read about them suffering a misfortune, they enjoyed a pleasurable spike in their brain's reward systems.

Similar findings have been revealed by researchers at Shenzhen University. Twenty-two participants were asked to play a simple computer game, then told (falsely) they were a 'two-star player'. Next, in a brain scanner, they were shown pictures of various 'one-star' and 'three-star' players receiving what looked to be painful facial injections. Afterwards, they claimed to have felt empathy for all the injectees. But their scans betrayed the lie: they only tended to experience empathy for the lower status 'one-star' players.

This was a small study, but consistent with other findings. Besides I'm not sure we really need neuroscientists to tell us that we struggle to empathise with higher-status people. We often feel all too comfortable mocking and bullying politicians, celebrities, CEOs and Prince Charles when, as hard as it can be to fathom, they're actually no less human than us.

Status play, like moral outrage, permeates human storytelling. It's hard to conceive of an effective story that doesn't rely on some form of status movement to squeeze our primal emotions, seize our attention, drive our hatred or earn our empathy. A study of over 200 popular nineteenth- and early twentieth-century novels found the antagonists' most common flaw was an ineffably chimpish 'quest for social dominance at the expense of others or an abuse of their existing power'.

Jane Austen was a master of such tales. When we meet 'handsome, clever and rich' Emma Woodhouse, we're motivated to keep reading by a desire to see her yanked down. Meanwhile, *Mansfield Park* tells of low-status Fanny Price whose struggling mother sends her to live with her wealthy uncle and aunt, Sir Thomas and Lady Bertram. Shortly before her arrival, as Lady Bertram is fretting that poor Fanny will 'tease' her 'poor pug', Sir Thomas girds himself to expect 'gross ignorance, some meanness of opinions and a very distressing vulgarity of manner'.

He's also concerned that she'll start thinking of herself as at one with her high-status cousins. Sir Thomas wishes for 'a distinction proper to be made between the girls as they grow up: how to preserve in the minds of my *daughters* the consciousness of what they are, without making them think too lowly of their cousin; and how, without depressing her spirits too far, to make her remember that she is not a *Miss Bertram*.' While he hopes his daughters will refrain from treating Fanny with arrogance, 'they cannot be equals. Their rank, fortune, rights and expectations will always be different.' If we weren't on Fanny's side before Sir Thomas's pronouncements, we are when we hear them. He's talking about us. We're Fanny Price. And we're fucking outraged.

3.8

William Shakespeare's *King Lear* shows what happens when humans undergo a nightmare even more dreadful than ostracisation. Shakespeare understood that there's nothing more likely to make a person mad, desperate and dangerous than the removal of their status. The play is a tragedy, a form that frequently shows how hubris – which can be viewed as the making of an unsound claim to status – can bring personal destruction. Such tales were told repeatedly by the Ancient Greeks and, of course, form real-life narratives that play out continually in chimp troops and human tribes. These dramatic status reversals have probably been part of our existence for millions of years.

King Lear is a canonical example of a story in which the

right external change strikes the right character at the right moment and thereby ignites a drama that feels as if it has its own explosive momentum. Its plot serves specifically to shatter its protagonist's deepest, most fiercely defended identity-forming beliefs. Just like the story of Charles Foster Kane, its ignition point and subsequent causes and effects are the seemingly inevitable consequences of its protagonist's flawed model of the world.

It all begins as an ageing Lear, heralded by trumpets, announces he'll divide his kingdom between his three daughters, its spoils being distributed in accordance to how well they perform in a love test. The more they adore him, the better the reward. In the defective reality that Lear's brain creates for him, he's the unrivalled, beloved and never-to-be-disputed king of everything around him. Lear naturally *accepts the reality of the world with which he's presented*. His neural models predict he'll consistently be treated with reverence and deference. This flawed model, which of course feels absolutely real and true, causes him to make mistakes that critically damage his ability to control the external world. When his manipulative daughters Regan and Goneril respond to his love test with extravagantly sycophantic oaths of boundless love, he doesn't question them. Why would he? They're simply reflecting the reality his brain's models are predicting. It would be like questioning the shining of the sun or the singing of the birds.

But Lear's third daughter, his favourite Cordelia, refuses to play. When she says she loves him no more or less than any daughter loves her father, she puts herself in conflict with his precious models. He responds as we all do, when our most sacred identity-forming beliefs are challenged. He pushes back. First, he threatens her: 'Mend your speech a little, lest it may mar your fortunes.' When she refuses, he disowns her: 'I disclaim all my paternal care.' Cordelia will now forever be 'a stranger to my heart and me'.

Lear's commitment to his flawed models is such that when the newly powerful Regan and Goneril begin conspiring to take everything from him, he struggles to perceive what's happening. As the predictions his models are making about the world increasingly fail, he reacts with denial, either in the form of ape-like rage or simple disbelief. When he discovers Goneril and her husband have put his messenger in the stocks, the insult is literally unbelievable to him. He's left sputtering and aghast. 'No, no, they would not . . . By Jupiter, I swear, no . . . They could not, would not do 't. 'Tis worse than murder to do upon respect such violent outrage.' When Goneril's assistant refers to him not as his 'King' but 'my Lady's father' he's overcome with fury – 'You whoreson dog! You slave! You cur!' – and physically attacks him.

When the reality of the external world finally becomes undeniable, Lear's internal model of it cracks apart. His entire self collapses. His theory of control had it that, to successfully manipulate his environment, all he had to do was issue orders. And this wasn't just a silly idea he could cast off when he realised it was false. It formed the very structure of his perception. It was the world he experienced as real. He saw evidence for its truth everywhere, and rubbished and denied any counter-knowledge, because that's exactly what brains do. It's from this sophisticated psychological understanding that the play gets its truth and drama. We can't simply toss aside our flawed ideas as if they're a pair of badly fitting trousers. It takes overwhelming evidence to convince us that 'reality' is wrong. When we finally realise something's up, breaking these beliefs apart means breaking ourselves apart. And that's precisely what happens in many of our most successful stories.

When Lear does break down, halfway through the play, it feels as if the entire planet's imploding. In an apocalyptic storm, he rages at skies, like a bleeding chimp brutally deposed by a conspiracy of younger animals. 'Here I stand, your slave, a poor, infirm, weak, and despised old man . . . I'll not weep. I have full cause of weeping, but this heart shall break into a hundred thousand flaws.' He's reduced to the position of beggar, this embodiment of the corrupt leader whose mistake was to forget that status, in human groups, should be earned.

Shakespeare knew well the psychological torments that can be unleashed by such a loss of status. In its most dangerous form, this is experienced as humiliation. In *Julius Caesar*, Cassius is at the heart of a conspiracy to kill the Roman leader who was once a friend. His hatred stems from an incident in childhood during which, on a dare, Cassius and Caesar tried to swim across the Tiber. But on this 'raw and gusty' day, Caesar failed. He was reduced to begging Cassius to save his life. His heroic act of costly signalling made, for Cassius, a model of the world in which he was forever superior in status to Caesar. But now they're grown up and that desperate, soggy boy has 'become a God, and Cassius is a wretched creature, and must bend his body if Caesar carelessly but nod on him'.

The rage that this unfair de-grading causes in Cassius is murderous.

Psychologists define humiliation as the removal of any ability to claim status. Severe humiliation has been described as 'an annihilation of the self'. It's thought to be a uniquely toxic state and is implicated in some of worst behaviours the human animal engages in, from serial murder to honour killings to genocide. In story, an experience of humiliation is often the origin of the antagonist's dark behaviour, whether it be murderous Cassius or *Gone Girl*'s scheming Amy Elliot Dunne, who could 'hear the tale, how everyone would love telling it' about how 'Amazing Amy' had been reduced to the level of those 'women whose entire personas are woven from a benign mediocrity' and about whom people think 'poor dumb bitch'.

Because humiliation is such an apocalyptic punishment, watching villains being punished this way can feel rapturous. As we're a tribal people with tribal brains, it doesn't count as humiliation unless other members of the tribe are aware of it. As Professor William Flesch writes, 'We may hate the villain, but our hatred is meaningless. We want him unmasked to people in his world.'

3.9

Babylon, 587 BC. A group of 4,000 high-status men and women were forced out of Jerusalem by King Nebuchadnezzar II. These Judeans journeyed long, and suffered, before finally finding a place to rest in the ancient city of Nippur. But they never forgot their beloved home. In exile, the Judeans determined to keep alive the customs of their people: their moral laws, their rituals, their language, their ways of living, eating and being. In order to do this it was essential that they preserved their stories.

Because most of these stories only existed orally, Judean scribes began writing them down on a series of scrolls. As they did, something remarkable happened. The ragbag of ancient myths and fables became connected. The scribes turned them into one complete cause-and-effect-laden tale. It began with the creation of the world and the first humans, Adam and Eve, and continued to include their occupation of Jerusalem.

The story had an astonishingly galvanising effect on this tribe of exiles. It acted as all tribal stories do, helping them function as a cooperative unit. As a list of prescribed behaviours, it enabled members to differentiate themselves from members of outside groups which created a psychological boundary between them and the 'other'. This same list of behaviours acted as a regulatory check-list against which they could police each other and therefore keep the tribe functional. But it also did much more. The story provided them with a heroic narrative of the world in which they were god's chosen people whose rightful homeland was Jerusalem. It filled the exiles with a sense of meaning, righteousness and destiny.

Seventy-one years after their banishment, the Judeans finally had the opportunity to return to their ancestral homelands. Led by a scribe named Ezra, they began their epic journey back to the glorious city they'd heard about only in stories. But when they finally arrived, they were horrified. The descendants of their low-status ancestors, who'd escaped the deportations, were rude, slovenly and interbreeding with other tribes. They weren't adhering to tribal laws about purity, food, worship or the Sabbath. Jerusalem itself was a crumbling mess.

For Ezra, such tribal decay was a catastrophe. He went to the temple, where it was believed their group's god Yahweh resided, and collapsed on the ground, wailing his despair and rage and betrayal. A crowd gathered. Ezra turned on them. They'd gravely offended Yahweh. They didn't deny it. But what could be done? He knew he had to somehow draw his people back together; to run into them the same tribal electricity that had held the exiles shoulder to shoulder, back in Babylon. There was only one way to do it: by unleashing the incredible power of their origin story.

Ezra had a wooden stage erected in a public place and sent out word something important was going to happen. A crowd formed. Ezra, flanked by twelve assistants, theatrically presented the scrolls on which their grand tribal narrative had been written. 'They immediately bowed their heads to the ground, as they would bow in the presence of their god, or their god's representative, in the temple,' writes Professor of English Martin Puchner. Something new was happening; something that would change the world forever. These scrolls, and the stories they contained, were being treated as if they themselves were sacred. And so a religion was born. 'Ezra's reading created Judaism as we know it.'

This might have been the first time a written story was treated as sacred, but human tribes have been bound together by such stories for tens of thousands of years. In our hunter-gatherer pasts much of our storytelling would've taken place around the campfire under the stars. Outrage and status-drenched tales of hunts and tribal exploits would've been told and retold, becoming ever-more magical and strange, eventually taking the form of sacred myth. Such stories would describe the nature of heroic behaviour. Certain characters would be celebrated, and gain status, for acting in ways the tribe approved of. Villainous or cowardly behaviour would trigger moral outrage – an urgent desire to see transgressors punished that would be satisfied in uproariously happy endings. In this way, stories transmitted the values of the tribe. They told listeners exactly how they ought to behave if they wanted to get along and get ahead in that particular group. There's a sense in which these stories would *become* the tribe. They'd represent what it stood for in ways purer and clearer than could any flawed human.

Stories are tribal propaganda. They control their group, manipulating its members into behaving in ways that benefit it. And it works. A recent study of eighteen hunter-gatherer tribes found almost eighty per cent of their stories contained lessons in how they should behave in their dealings with other people. The groups with the greater proportion of storytellers showed the most pro-social behaviour.

Because one of our deepest and most powerful urges is the gaining of ever more status, our tribal stories tell us how to earn it. A human tribe can be viewed as a status game that all its members are playing, its rules being recorded in its stories. Every human group that has a shared purpose is held together by such stories. A nation has a story it tells about itself, in which its values are encoded, as does a corporation and a religion and a mafia organisation and a political ideology and a cult. The Bible, The Qur'an and the Torah that Ezra presented to his people in Jerusalem are ready-made theories of control that are internalised by their followers, instructing them how to behave in order to achieve connection and status.

Some of our oldest recorded stories transmit such rules. *The Epic of Gilgamesh*, which pre-dates Ezra's story by more than a thousand years and even lends it its episode about a worldwide flood, tells of a King who, like Shakespeare's Lear, has forgotten that status should be earned. In its first section, the gods send down a challenger, Enkidu, to humble him. King Gilgamesh and Enkidu become friends. Together they bravely take on the monster of the forest, Humbaba, using superhuman effort to slay him before triumphantly returning with valuable wood to continue building Gilgamesh's great city. By the end of the saga, Enkidu has died, but King Gilgamesh is fully humbled, accepting his lot as just another mortal human. We think more of him and thereby reward him with a bump in status.

That 4,000-year-old epic provides the same tribal function as *Mr Nosey*. In Roger Hargeaves's children's book, the protagonist's flawed model of the world tells him he'll only be safe if he sticks his long nose into other people's business. But the villagers plot against him, first daubing paint on his prying nose, then banging it with a hammer. Finally humbled, Nosey mends his ways, 'and soon became friends with everybody in Tiddletown'. For shedding his anti-social habits, Nosey is rewarded with connection and status.

All of us are being silently controlled by any number of instructional stories at once. A unique quality of humans is that we've evolved the ability to *think* our way into many tribes simultaneously. 'We all belong to multiple in-groups,' writes Professor Leonard Mlodinow. 'As a result our self-identification shifts from situation to situation. At different times the same person might think of herself as a woman, an executive, a Disney employee, a Brazilian or a mother, depending on which is relevant – or which makes her feel good at the time.'

These groups, and their stories of how to behave and gain connection and status, form part of our identity. It's mostly during adolescence, that period in which we're composing our 'grand narrative of self', that we decide which 'peer groups' to join. We seek out people who have similar mental models to us – who have comparable personalities and interests and perceive the world in ways we recognise. Late adolescence sees many choosing a political ideology, left or right – a tribal master-story that fits over our unconscious landscape of feelings and instincts and half-formed suspicions and makes sense of it, suddenly infusing us with a sense of clarity, mission, righteousness and relief. When this happens it can feel as if we've encountered revealed truth and our eyes have suddenly been opened. In fact, the opposite has happened. Tribal stories blind us. They allow us to see only half the truth, at best.

The psychologist Professor Jonathan Haidt has explored the stories that competing ideological tribes tell about the world. Take capitalism. For the left, it's exploitative. The Industrial Revolution gave evil capitalists the technology to use and abuse workers as dumb machine-parts in their factories and mines and reap all the profits. The workers fought back, unionising and electing more enlightened politicians and then, in the 1980s, the capitalists became resurgent, heralding an era of ever-increasing inequality and eco-disaster. For the right, capitalism is liberation. It freed the used and abused workers from exploitation by kings and tyrants and gave them property rights, the rule of law and free markets, motivating them to work and create. And yet this great freedom is under constant attack from leftists who resent the idea that the most productive individuals are properly rewarded for their hard work. They want everyone to be 'equal and equally poor'.

What's insidious about these stories is that they each tell only a partial truth. Capitalism is liberating and it's also exploitative. Like any complex system it has a trade-off of effects, some good, others bad. But thinking with tribal stories means shutting out such morally unsatisfying complexity. Our storytelling brains transform reality's chaos into a simple narrative of cause and effect that reassures us that our biased models, and the instincts and emotions they generate, are virtuous and right. And this means casting the opposing tribe into the role of villain.

The evil truth about humans is that we don't just compete for status with other people *inside* our tribes. The tribes we belong to also compete with rival tribes. We're not harmlessly groupish like starlings or sheep or shoals of mackerel, but violently so. In the twentieth century alone, tribal conflict killed 160 million, whether by genocide, political oppression or war.

We have this in common with the chimpanzee whose males, sometimes accompanied by females, patrol the boundaries of their territory, halting in silence for as long as an hour to listen for enemy movements. When caught, a 'foreign' chimp is savagely beaten to death: arms twisted off, throat torn out and fingernails plucked, genitals ripped off, the warriors gulping down the gushing blood. When all the males of a neighbouring troop are killed or chased out, the victorious chimps take over their territory and the females still in it. The primatologist Professor Frans de Waal writes that 'it cannot be coincidental that the only animals in which gangs of males expand their territory by deliberately exterminating neighbouring males happen to be humans and chimpanzees. What is the chance of such tendencies evolving independently in two closely related mammals?'

We still have this primitive cognition. We think in tribal stories. It's our original sin. Whenever we sense the status of our tribe is threatened by another, these foul networks fire up. In that moment, to the subconscious brain, we're back in the prehistoric forest or savannah. The storytelling brain enters a state of war. It assigns the opposing group purely selfish motives. It hears their most powerful arguments in a particular mode of spiteful lawyerliness, seeking to misrepresent or discard what they have to say. It uses the most appalling transgressions of their very worst members as a brush to smear them all. It takes its individuals and erases their depth and diversity. It turns them into outlines; morphs their tribe into a herd of silhouettes. It denies those silhouettes the empathy, humanity and patient understanding that it lavishes on its own. And, when it does all this, it makes us feel great, as if we're the moral hero of an exhilarating story.

The brain enters this war state because a psychological tribal threat is a threat to its theory of control – its intricate network of millions of beliefs about how one thing causes another. Its theory of control tells it, among many other things, how to get what it most desires, namely connection and status. It forms the scaffolding of the model of the world and self it has been building since birth.

Of course this model, and its theory of control, is indivisible from who we are. It's what we're experiencing, in the black vault of our skulls, as reality itself. It's hardly surprising we'll fight to defend it. Because different tribes live by different models of control – communists and capitalists, to take a broad example, award their prizes of status and connection for very different behaviours – a tribal challenge is existentially disturbing. It's not merely a threat to our surface beliefs about this and that, but to the very subconscious structures by which we experience reality.

It's also a threat to the status game to which we've invested the efforts of our lives. To our subconscious, if another tribe is allowed to win, their victory won't merely pull us down the hierarchy but will destroy the hierarchy completely. Our loss in status will be complete and irreversible. This removal of the ability to claim status meets the psychologist's definition of humiliation, that 'annihilation of the self' which underlies a saturnine suite of murderous behaviours, from spree shootings to honour killings. When a group's collective status feels threatened and they fear even the possibility of humiliation by another group, the result can be massacre, crusade and genocide. Such dynamics have played out relatively recently in places such as Rwanda, the Soviet Union, China, Germany, Myanmar, the southern states of America and, of course, Ezra's precious Jerusalem.

In such times, tribes deploy the explosive power of story, with all its moral outrage and status play, in order to galvanise and motivate their members against the enemy. In 1915, the film *The Birth of a Nation* presented African Americans as unintelligent brutes who sexually bullied white women. The three-hour-long story played to sold-out crowds and recruited thousands to the Ku Klux Klan. In 1940, one year before the release of *Citizen Kane*, the film *Jew Süss* portrayed Ezra's descendants as corrupt and showed a high-status Jewish banker, Süss Oppenheimer, raping a blonde German woman, before being hanged in front of grateful crowds in an iron cage. It premiered at the Venice Film Festival, winning the Golden Lion Award, was seen by twenty million and caused viewers to pour en masse into the streets of Berlin chanting, 'Throw the last of the Jews out of Germany.' That sexual violence against females appeared in both films and is a territorial dominance behaviour of chimpanzees is surely no coincidence.

But such stories don't only exploit outrage and tribal humiliation for their power. Many deploy a third incendiary group emotion: disgust. In our evolutionary pasts, the threat from competing groups wouldn't come only from their potential for violence. They could also be carrying dangerous pathogens that our immune systems hadn't previously encountered and so couldn't defend us against. Exposure to carriers of pathogens – in faeces, say, or rotten food – naturally activates feelings of disgust and revulsion. Our tribal brains seem to have developed the cultural tic of thinking of foreign tribes in such a way. This, perhaps, is why children still commonly hold their noses as a way of derogating members of out-groups.

Tribal propaganda exploits these processes by representing enemies as disease-carrying pests such as cockroaches, rats or lice. In *Jew Süss*, the Jewish people are portrayed as filthy and unhygienic and are shown teeming into a city as a plague. Even popular conventional stories exploit the power of disgust. Villains from *Harry Potter*'s Lord Voldemort to *Beowulf*'s Grendel to *The Texas Chainsaw Massacre*'s Leatherface have disfigurements that fire these neural networks. In *The Twits*, Roald Dahl created a typically marvellous confabulation of the disgust principle: 'If a person has ugly thoughts, it begins to show on the face. And when that person has ugly thoughts every day, every week, every year, the face gets uglier and uglier until it gets so ugly you can hardly bear to look at it.'

It's in these ways that story both exposes and enables the worst traits of our species. We willingly allow highly simplistic narratives to deceive us, gleefully accepting as truth any tale that casts us as the moral hero and the other as the two-dimensional villain. We can tell when we're under its power. When *all* the good is on our side and *all* the bad on theirs, our storytelling brain is working its grim magic in full. We're being sold a story. Reality is rarely so simple. Such stories are seductive because our hero-making cognition is determined to convince us of our moral worth. They justify our primitive tribal impulses and seduce us into believing that, even in our hatred, we are holy.

3.10

It's sometimes assumed that we root for characters who are simply kind. This is a nice idea, but it's not true. In story, as in life, kind people are wonderful and inspiring and oh so terribly boring. Besides, if a hero starts out in such perfect selfless shape there's going to be no tale to tell. For the story theorist Professor Bruno Bettelheim, the storyteller's challenge isn't so much one of arousing the reader's moral respect for the protagonist, but their sympathy. In his inquiry into the psychology of fairy tales, he writes that 'the child identifies with the good hero not because of his goodness, but because the hero's condition makes a deep positive appeal to him. The question for the child is not, "Do I want to be good?" but "Who do I want to be like?"'

But if Bettelheim is correct, how do we explain antiheroes? Millions have been entranced by the adventures of Humbert Humbert, the protagonist of Vladimir Nabokov's *Lolita*, who embarks on a sexual relationship with a twelve-year-old girl. Surely we don't want to be 'like' him?

In order to achieve his trick of not having us throw his novel into a cleansing fire after the first seven pages, Nabokov has to go to sometimes extreme lengths to subconsciously manipulate our tribal social emotions. In a scholarly introduction written by an academic we immediately learn that Humbert is dead. Next we discover that, prior to his passing, he was in 'legal captivity' awaiting trial. This immediately deflates much of our moral outrage before we even get the chance to feel it: the poor bastard's caught and dead. Whatever he's done, he's had his tribal comeuppance. We can relax. The craving subsides. Before the first sentence is even finished, Nabokov has begun slyly freeing us to enjoy what's to come.

When we meet the man himself our outrage is further punctured by his immediate acknowledgements of wrong-doing, calling Lolita 'my sin' and himself a 'murderer'. It helps, too, that Humbert's the opposite of disgusting, being handsome, well-tailored and charming. He's darkly funny, dealing with the death of his mother in perhaps the most famous in-parenthesis aside in literature – '(picnic, light-ning)' – and describing Lolita's mother as looking like 'a weak solution of Marlene Dietrich'. We learn his hebephiliac tendencies were triggered by tragedy: when he himself was twelve, his first love Annabelle died, 'that little girl with her seaside limbs and ardent tongue haunted me ever since – until at last, twenty-four years later, I broke her spell by incarnating her in another.'

When Humbert's adult interest in girls of Annabelle's age becomes apparent, he tries to cure himself with therapy and marriage. It doesn't work. The story's ignition point (just as it is for Charles Foster Kane and King Lear) is an inevitable consequence of his flawed model of the world: Humbert meets and falls in love with Lolita. We soon realise the girl's mother despises her: not only has she given her daughter 'the meanest and coldest' room in the house, Humbert finds a personality questionnaire she's filled in on her behalf. It indicates she believes Lolita to be, 'aggressive, boisterous, critical, distrustful, impatient, irritable, negativistic (underlined twice) and obstinate. She had ignored the thirty remaining adjectives, among which were cheerful, cooperative, energetic, and so forth. It was really maddening.' She then packs her off, against her will, to a boarding school with 'strict discipline'. By a variety of powerful and crafty means, Nabokov is manipulating our emotions such that we find ourselves somewhat rooting for Humbert.

If Humbert is to have Lolita, her mother has to go. Will Humbert kill her? Nabokov knows he's already asking a great deal from the reader. Our social emotions are only on Humbert's 'side' in the most fragile way and certainly won't stand watching him kill. So when her death takes place, it's not directly Humbert's doing. In perhaps his most audacious piece of manipulation, Nabokov has his protagonist unable to bring himself to commit the awful deed. Instead he relies on what he cheekily has Humbert describe as 'the long hairy arm of coincidence' to do it for him. She's run over by a car.

When Humbert finally gets his hands on Lolita, he's randy but also conflicted, hesitant and guilty. We crucially discover she's no longer a virgin, having already slept with a boy at summer camp. She's presented, at least by our unreliable narrator, as unsympathetic – pushy, confident, manipulative and precocious – and because this is the behaviour we're *shown*, it's what we'll subconsciously and emotionally respond to. Lolita comes to dominate Humbert before deciding to run off with a far more despicable man, Clare Quilty. Where Nabokov sympathetically manipulates our response to Humbert, he fully unleashes the disgust principle against this 'subhuman' predatory hebephiliac pornographer: we see the 'black hairs on the back of his piggy hands' and watch him 'scratching loudly his fleshy and gritty grey cheek and showing his small pearly teeth in a crooked grin'. Then, in a thrilling act of altruistic punishment and costly signalling that we're by now deeply craving, Humbert kills him.

Our antihero finally departs the story having submitted voluntarily to his arrest. The very last thing he shares with us is a confessional memory from the period following his abandonment by Lolita. He'd pulled up in his car at the side of a high valley, at the bottom of which lay a small mining town. In its streets, he heard the voices of playing children: 'I stood listening to that musical vibration from my lofty slope, to those flashes of separate cries with a kind of demure murmur for background, and then I knew that the hopelessly poignant thing was not Lolita's absence from my side, but the absence of her voice from that concord.' Humbert Humbert might have done a terrible thing, but Nabokov's ability to manipulate our deepest tribal feelings about his sin, his soul, are tremendous.

Similar manipulations take place on behalf of other antiheroes, not least the protagonist of the television series *The Sopranos*. Our first meeting with the Mafioso Tony Soprano occurs in a psychotherapist's waiting room. We learn he developed a bond with some ducks and ducklings that regularly landed in his pool, and suffered a panic attack when they finally left. He weeps when he speaks of them. Not only is

Soprano sensitive and in pain, he's relatively low in status. Far from being some all-powerful John Gotti, he's the capo of a marginal New Jersey gang and, anyway, as he says to his new therapist: 'I came in at the end, the best is over.'

When we see Soprano beating a man, the victim is just a 'degenerate fucking gambler' who owes him money and insulted him: 'you've been telling people I'm nothing compared to the people who used to run things.' As the episode unfolds, Soprano secretly tries to help a non-mob friend in whose restaurant his much more horrible uncle has planned a hit. Soprano cares for his mother. When he takes her to a prospective nursing home and she becomes distressed, he suffers another anxiety attack. We then discover she's plotting with his uncle to have him killed.

The author Patricia Highsmith indulges in similar manipulations. In *Ripley's Game*, the sociopathic con artist Tom Ripley is handsome, eloquent and cultured, just like Humbert. And, like Humbert and Soprano, he's in conflict with a much more evil villain, Reeves Minot. Like Soprano, darker, more powerful forces are ranged against him in the form of the Italian mafia. And so on. If we have the alarming realisation that we're actually rooting for these characters, it's because we're being cleverly manipulated into doing so by everything that's happening around them. They might be sex criminals, con artists and gangsters, but the world that's created for them to battle against is such that we overlook their deviancies in spite of ourselves.

There's a sense in which all protagonists are antiheroes. Most, when we meet them, are flawed and partial and only become truly heroic if and when they manage to change. Any attempt to find a single reason why we find characters root-worthy is probably destined to fail. There isn't one secret to creating empathy but many. The key lies in the neural networks. Stories work on multiple evolved systems in the brain and a skilled storyteller activates these networks like the conductor of an orchestra, a little trill of moral outrage here, a fanfare of status play over there, a tintinnabulation of tribal identification, a rumble of threatening antagonism, a tantara of wit, a parp of sexual allure, a crescendo of unfair trouble, a warping and wefting hum as the dramatic question is posed and reposed in new and interesting ways – all instruments by which masses of brains can be captivated and manipulated.

But I suspect there's also something else going on. Story is a form of play that we domesticated animals use to learn how to control the social world. Archetypal stories about anti-heroes often end in their being killed or otherwise humiliated, thus serving their purpose as tribal propaganda. We're taught the appropriate lesson and left in no doubt about the costs of such selfish behaviour. But the awkward fact remains that, as we experience the story unfolding in our minds, we seem to enjoy 'playing' the antihero. I wonder if this is because, somewhere in the sewers far beneath our hero-making narrators, we know we're not so lovely. Keeping the secret *of* ourselves *from* ourselves can be exhausting. This, perhaps, is the subversive truth of stories about antiheroes. Being freed to be evil, if only in our minds, can be such a joyful relief.

3.11

If Joseph Campbell is correct in saying the only way of describing a human 'truly' is by describing their imperfections, how might a storyteller describe you? That is, what are the identity-forming beliefs you cling to and define you that are wrong and often harm you? What's your version of the butler Stevens's emotional restraint? This, in all likelihood, will not be a straightforward question to answer. The reason such flaws are pernicious is that they're often invisible to us. They're a component part of our controlled hallucination of reality. Worse, when they do become noticeable, our sly hero-making brains work to make them seem as if they're not flaws at all, but virtues. We fight to defend them.

For me, I suspect it might be a foundational belief that other people are dangerous. I have a theory of control that says in order to remain safe humans should be avoided wherever possible. As I've grown older, and more adept at being social, I've developed a range of selves I wear in public like masks so I can function. But I've also retreated further into myself. My crafty brain tells me the decisions that have brought me to where I am have been brilliant. It's wonderful I live such a relatively peaceful life, cocooned in the country with my wife and dogs. *'Hell is other people, yeah?'*

But, sometimes, I'm not sure. In my thirties, I'd occasionally pine for friends but whenever I got the chance to make any, would pull away. I no longer pine. As the world has quietened around me, I've come to know pleasant solitude and bitter loneliness as two expressions of the same face. One can become the other in a flash. I feel myself becoming odder. I sense it in the wary eyes of the postman and people I meet when out walking. I worry about my wife and I getting old, childless and isolated. But what to do? My neural models have been organising themselves in this direction for decades. In order to break them apart, and change the flawed, core belief they're founded upon, something properly *dramatic* would have to happen.

I can tell a convincing tale of how my flawed model of the world came to be. Tests on the big five personality scale have me low in extraversion and high in neuroticism. These genetic tendencies were exacerbated by a difficult childhood home-life. I tried to find what I was missing at school but my desperation just alienated and irritated. Then alcohol happened, and drugs. Giving them up meant not socialising and that turned out to be surprisingly seductive. What I'd always wanted was noise and people. What I'd needed, it seemed, was the opposite. That's the neat, neocortical cause-and-effect origin story I have of myself.

Is this confabulation true? Probably, some of it. How much I'll never know. I'm sure we all have such stories though. In our therapeutic age we're conditioned to seek soothing just-so stories, in our pasts, that explain the origin of our damage. Although this seems to have become more habitual in twentieth-century fiction, it's been happening for centuries, not least in Shakespeare's account of the origin of Cassius's psychological damage in *Julius Caesar* – his murderous obsession coming into being in his youth in the choppy waters of the river Tiber.

The story of *Citizen Kane* is itself a hunt for origin damage.

Rawlston's newsmen are charged to discover who this man was, who'd inherited a spectacular fortune and yet chosen to run a newspaper, attempted to go into politics and then died alone and unhappy in 'the world's largest pleasure ground' surrounded by a 'collection of everything so big it can never be catalogued'. Most specifically, they're sent to uncover the mystery of the last word he ever spoke: rosebud.

During the search, one of Rawlston's men reads the memoir of the guardian who'd raised Kane from boyhood. Its pages reveal that his mother gave Kane up to a wealthy guardian, Thatcher, against his father's will. She believed she was doing the right thing because his father beat him. But Charles's father also believed *he* was doing the right thing, his hero-making narrator insisting the beatings were for the boy's own good. Despite the corporal punishment, young Charles was essentially happy. We're shown him full of life, joyfully playing soldiers in the snow. When Thatcher takes him away, he attacks him with a sledge.

In the film's final frames, the information gaps that opened-up at the story's start are finally closed. We discover that, written on that sledge was the word 'rosebud'. The glass snow globe Kane dropped and smashed when he died contained a house resembling that of his parents. In being wrenched away from that home, a void was created that he spent his life trying to fill with the love of the masses and all the material possessions he could buy. But the hole was too big. It was during that moment with the sledge that the damage took place to his models that, in turn, created the ignition point and plot of his story. This revelation answers the fundamental dramatic question of *who is he?* and thereby leaves the viewer moved and satisfied.

The origin damage suffered by the butler Stevens took place against the background of his childhood. He was raised to believe in Britain's greatness when the nation was still inarguably powerful. But there was one moment, in his past, that seems especially formative. He hears a story about his father, a head butler, and how he dealt with a particular visitor he 'detested' – an army General whose unprincipled and irresponsible actions during the Second Boer War led directly to the death of his eldest son, and our narrator's brother. When the General arrived at the house without a valet, Stevens Senior volunteered to look after him. As he tended to him in 'intimate proximity', throughout his four-day visit, the General proved arrogant and rude. But even as he boasted of his military accomplishments, Stevens Senior betrayed not a twitch of the turbulent emotions he had to endure. This dignity in emotional restraint became an idealised model of self, in Stevens's mind. It was incorporated into his theory of control. The story told him who he had to be in order to be welcomed into the status game of butlers and climb to its pinnacle.

It seems characteristic of many successful stories that their authors reduce origin damage to specific moments. It doesn't do to be general and say, for example, 'it's because their parents didn't love them enough,' because such vague thinking can only lead to more vague thinking. In reality, of course, origin damage is often a matter of grim erosion, commonly taking place over months, years and repeated bloody incidents. But it's my experience when teaching these principles that, if we're creating stories, specificity is essential. It helps to pin a character's damage down to an actual event and imagine it thoroughly, even if these scenes are left out of the play, screenplay or novel that follows, as they often are. It's only once the writer knows when it happened, how it happened and what flaw the incident created, they can begin to truly know their character. That belief comes to define them. Their self-reinforcing brains begin to see evidence supportive of it everywhere.

When you've broken a character you can begin to build their story. And they should be broken in a way that's specific. Ishiguro's butler's mistake was specifically emotional restraint. That's the grain around which his entire life, and the novel that tells of it, comes alive.

If origin damage in story most often occurs in youth, it's because it's in the first two decades of life that we're busy forming ourselves out of our experiences. It's when our models of reality are being built. (If you want to imagine how bizarre and berserk a person with unbuilt neural models of reality would be like, just imagine a four-year-old.) (Or a fourteen-year-old.) As adults, the hallucination we experience as truth is built out of our pasts. We see and feel and explain the world partly with our damage.

This damage can take place before we're even able to speak. Because humans crave control, infants whose caregivers behave unpredictably can grow up in a constant state of anxious high alert. Their distress gets built into their core concepts about people which can lead to significant social problems when they're grown. Even a lack of affectionate touch, in our earliest years, has the potential to hurt us forever.

The body has a dedicated network of touch receptors optimised specifically to respond to being stroked. For the neuroscientist Professor Francis McGlone, gentle stroking is critical for healthy psychological development. 'My hunch is that the natural interaction between parents and the infant – that continuous desire to touch, cuddle and handle – is providing the essential inputs that lay the foundations for a well-adjusted social brain,' he has said. 'It's more than just nice, it's absolutely critical.'

The brain's models continue to form during adolescence. Our popularity or otherwise at school also warps our neural models, and therefore our experience of reality, forever. Our position on the social hierarchy during adolescence doesn't merely alter who we are as adults superficially, writes the psychologist Professor Mitch Prinstein, it changes 'our brain wiring and, consequently, it has changed what we see, what we think and how we act'.

Researchers asked people to watch videos of scenes that were busy with social interactions, such as film of a school corridor. They then tracked their saccades so they could see which elements the participant's brains were attending to. Those with 'past histories of social success' spent most of their time on people being friendly – smiling, chatting, nodding. But those who'd had high-school experiences of loneliness and social isolation 'scarcely looked at the positive scenes at all', writes Prinstein. Instead they spent around eighty per cent of their time looking at people being unfriendly and bullying. 'It was as if they had watched a completely different movie.'

Similar tests had people viewing simple animations of shapes interacting in ambiguous ways. Participants who'd been unpopular at school tended to tell a spontaneous cause-and-effect story about what was happening in which the shapes were behaving violently towards each other. Those who'd been popular were much more likely to experience them as joyfully playing.

This is how we go through our day-to-day lives. What we see in our human environments is a product of our pasts – and, all too often, a product of our own particular damage. We're literally blind to that which the brain ignores. If it sends the eye to only the distressing elements around us, that's all we'll see. If it spins cause-and-effect tales of violence and threat and prejudice about actually harmless events, that's what we'll experience. This is how the hallucinated reality in which we live at the centre can be dramatically different to that of the person we're standing right next to. We all exist in different worlds. And whether that world feels friendly or hostile depends, in significant part, on what happened to us as children. 'At some level, without our being aware of it, our brains spend all day, every day, drawing upon initial, formative high school memories.'

Harmful childhood experiences damage our ability to control the environment of other people. And for us domesticated creatures the environment of other people is everything. All principal characters in story will engage in such struggles. It might seem as if certain kinds of fictional stories don't concern such characters – Indiana Jones or heroes of boys-own war adventures, such as Andy McNab's *Bravo Two*

Zero, for example, focus on a protagonist's attempts at controlling the physical world rather than the social. But even they will ultimately have to grapple with antagonistic minds, whether in the form of some villain or their own tumultuous, rebelling subconscious.

Because origin damage happens when our models are still being built, the flaws it creates become incorporated into who we are. They're internalised. The self-justifying hero-maker narrative then gets to work telling us we're not partial or mistaken at all – we're right. We see evidence to support this false belief everywhere, and we deny, forget or dismiss any counter-evidence. Experience after experience seems to confirm our rightness. We grow up looking out of this broken model of the world that feels absolutely clear and real, despite its warps and fissures.

Every now and then, actual reality will push back at us. Something in our environment will change in such a way that our flawed models aren't predicting and are, therefore, specifically unable to cope with. We try to contain the chaos but because this change strikes directly against our model's particular flaws, we fail. Then we can become conflicted. Are we right? Or is there actually a chance we're wrong? If this deep, identity-forming belief turns out to be wrong, then who the hell are we? The dramatic question has been triggered. The story has begun.

Finding out who we are, and who we need to become, means accepting the challenge that story offers us. Are we brave enough to change? Can we become a hero?

This is the question a plot, and a life, asks of each us.

CHAPTER FOUR:

PLOTS, ENDINGS
AND MEANING

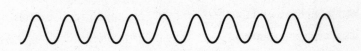

GRIFFITH COLLEGE DUBLIN
SOUTH CIRCULAR ROAD DUBLIN 8.
Tel: 01 4150490 Fax: (01) 4549265
library@griffith.ie

4.0

A hero is selfless. A hero is courageous. A hero earns status. But heroes, in story and in life, have a final essential quality that we've yet to fully encounter. This is our oldest and most fundamental drive, probably originating back to when we were single-celled organisms. Humans are directed towards goals. We want things and we strive to get them. When unexpected change strikes we don't just climb back into bed and hope it all goes away. Well, we might for a while. But at some point we stand up. We face it. We *fight*. For the nineteenth-century critic Ferdinand Brunetière this was the one inviolable rule of drama: 'What we ask of the theatre is the spectacle of a will striving towards a goal.' Fundamental to successful stories and successful lives is the fact that we don't passively endure the chaos that erupts around us. These events challenge us. They generate a desire. This desire makes us act. This is how change summons us into the adventure of the story, and how an ignition point sprouts a plot.

Goal-direction is the foundational mechanism on top of which all our other urges are built. The basic Darwinian aim of all life forms is to survive and reproduce. Because of the peculiarities of our evolutionary history, human strategies to attain these goals centre on achieving connection with tribes, and on status within them. On top of these deep universals sits everything else we desire – our ambitions, feuds, love affairs, disappointments and betrayals. All of our struggles. All the stuff of story.

Humans have a compulsion to make things happen in their environment that's so powerful it's described by psychologists as 'almost as basic a need as food and water.' When researchers put people in flotation tanks and block their eyes and ears they find that, often within seconds, they'll start rubbing their fingers together or making ripples in the water. After four hours some are singing 'bawdy songs'. Another study found 67 per cent of male participants and 25 per cent of female participants so desperate to make things happen in a room that was empty of stimulus, except for an electric-shock machine, that they started giving themselves painful shocks. Humans do things. They act. We can't help it.

Our goals give our lives order, momentum and logic. They provide our hallucination of reality with a centre of narrative gravity. Our perception organises itself around them. What we see and feel, at any given moment, depends on what we're trying to get – when we're caught in the street in a downpour of rain, we don't see shops and trees and doorways and awnings, we see places of shelter. Goal-direction is so important to human cognition that when information about it is absent we can enter a state of bafflement. The psychologists Professors John Bransford and Marcia Johnson asked people to remember the following passage:

The procedure is actually quite simple. First you arrange things into different groups depending on their makeup. Of course, one pile may be sufficient depending on how much there is to do. If you have to go somewhere else due to lack of facilities that is the next step, otherwise you are pretty well set. It is important not to overdo any particular endeavour. That is, it is better to do too few things at once than too many. In the short run this may not seem important, but complications from doing too many can easily arise. A mistake can be expensive as well. The manipulation of the appropriate mechanisms should be self-explanatory, and we need not dwell on it here. At first the whole procedure will seem complicated. Soon, however, it will become just another facet of life. It is difficult to foresee any end to the necessity for this task in the immediate future, but then one never can tell.

Most failed to recall more than a handful of sentences. But a second group were told prior to reading that the paragraph concerned the washing of clothes. The simple addition of a human goal transformed the gobbledegook into something clear. They remembered twice as much.

In order to encourage us to act, to struggle, to *live*, the hero-making brain wants us to feel as if we're constantly moving towards something better. Assuming we're mentally healthy, we're pushed on into our plots by a delusional sense of optimism and destiny. One clever study asked restaurant employees to circle all the likely possibilities for their own future lives, before doing the same on behalf of a liked colleague. Many more circles appeared for their lives than for their co-workers. Another test found that eight in every ten participants believed things would turn out better for them than for others.

Goal-direction helps give story its thrill. As the protagonist pursues their goal we *feel* their struggle. As they grab for their prize, we experience their joy. As they fail, we cry out. In life and in story our emotions tell us what's of value. Our emotions guide us, letting us know who we ought to be and what we should go after, using a language millions of years older than words. When we're behaving heroically, we *feel* we're doing so because our actions are being soundtracked by positive emotions. Humans are by no means unique in this. The psychologist Professor Daniel Nettle writes that 'when an amoeba follows a chemical gradient to reach and then ingest some food, we might say that it is acting on its positive emotions. All sensate organisms have some kind of system for finding good things in the environment and going after them, and the suite of human positive emotions is just a highly developed system of this kind.'

Video games plug directly into such core desires. Multiplayer online games, such as *World of Warcraft* and *Fortnite*, are stories. When a player logs on and teams up with fellow players to embark on a difficult mission, their three deepest evolved cravings are powerfully fed – they experience connection, earn status and are given a goal to pursue. They become an archetypal hero battling through a three-act narrative of crisis-struggle-resolution. Modern games are so ferociously effective at feeding these human fundamentals that they can become addictive, with 'gaming disorder' now classified as a disease by the World Health Organisation. One Welsh teenager, Jamie Callis, would spend up to twenty-one hours per day playing *Runescape*. 'One minute you'd be chopping trees and the next you'd be killing something or going on a quest,' he told his local newspaper. 'You had clans of people, and that's where you'd really have a family.' Callis spent so much time conversing with his American and Canadian teammates that he began losing his Welsh accent. In South Korea, two parents became so overwhelmingly engrossed in a multiplayer game that they allowed their three-month-old daughter to starve to death. The game that obsessed them, *Prius Online*, partly involved nurturing and forming an emotional bond with 'Anima', a virtual girl.

The psychologist Professor Brian Little has spent decades studying the goals that humans pursue in their everyday lives. He finds we have an average of fifteen 'personal projects' going at once, a mixture of 'trivial pursuits and magnificent obsessions'. These projects are so central to our identity that Little likes to tell his students, 'We *are* our personal projects.' His studies have found that, in order to bring us happiness, a project should be personally meaningful and we ought to have some level of control over it. When I asked him if a person pursuing one of these 'core' projects was a bit like an archetypal hero battling through a three-act narrative of crisis-struggle-resolution he said, 'Yes. A thousand times yes.'

Little isn't the first to argue that the fundamental human value is the struggle towards a meaningful goal. In Ancient Greece, Aristotle tried to puzzle out the true nature of human happiness. Some posited a 'hedonic' form defined by pleasure and the satisfaction of short-term desires. But Aristotle contemptuously dismissed the hedonists, saying that, 'The life they decide on is a life for grazing animals.' Instead, he described the idea of 'eudaemonia'. For Aristotle happiness was not a feeling but a practice. 'It's living in a way that fulfils our purpose,' the classicist Professor Helen Morales said. 'It's flourishing. Aristotle was saying, "Stop hoping for happiness tomorrow. Happiness is being engaged in the process."'

Recent extraordinary evidence that humans are built to live according to Aristotle's concept of happiness as a practice rather than a goal comes from the field of social genomics. Results from a team led by Professor of medicine Steve Cole suggest health can improve – risk of heart disease, cancer and neurodegenerative disorders going down; antiviral response going up – when we're high in eudaemonic happiness. It changes the expression of our genes. Studies elsewhere find that living with a sufficient sense of purpose reduces the risk of depression and strokes and helps addicts recover from addiction. People more likely to agree with statements such as, 'Some people wander aimlessly through life, but I am not one of them,' have been found to live longer, even when other factors are controlled for.

When I asked Cole to define eudaemonia he said it was 'kind of striving after a noble goal'.

'So it's heroic behaviour in a literary sense?'

'Right,' he said. 'Exactly.'

Humans are built for story. When we push ourselves towards a tough yet meaningful goal, we thrive. Our reward systems spike not when we achieve what we're after but when we're in pursuit of it. It's the pursuit that makes a life and the pursuit that makes a plot. Without a goal to follow and at least some sense we're getting closer to it, there is only disappointment, depression and despair. A living death.

When a threatening and unexpected change strikes, our goal is to deal with it. This goal possesses us. The world narrows. We enter a kind of cognitive tunnel and see only our mission. Everything in front of us becomes either a tool to help us achieve our desire or an obstacle we must kick aside. This is also true for protagonists in story. Without Brunetière's *will striving towards a goal* being present in the scene of a story, there's no drama, only description.

This narrowing should be especially present at a story's ignition point. But this is exactly where many stories fail. In order to be maximally compelling, protagonists should be active, the principal causer of effects in the plot that follows. Textual analyses reveal the words 'do', 'need' and 'want' appear twice as often in novels that feature in the *New York Times* bestseller list as those that don't. A character in a drama who isn't reacting, making decisions, choosing and trying somehow to impose control on the chaos isn't truly a protagonist. Without action, the answer to the dramatic question never really changes. Who they are is who they always were, but slowly, dully sinking.

4.1

What happens next? This is the question brilliant scholars from Aristotle onwards have spent centuries attempting to answer. What does the protagonist have to do in order to lead reader and viewer on a maximally satisfying story? The pursuit of the perfect plot traditionally involves theorists gathering a number of successful myths and tales together and running their divining rods over them in an attempt to detect their hidden patterns. Their findings have been hugely influential. They shape today's landscape of popular storytelling.

For the mythologist Joseph Campbell, a story starts with a hero receiving and, at first, refusing a call to adventure. A mentor then comes along to help them change their mind, then into the plot they go. Somewhere in the middle, they'll cross a transformational threshold, only to rouse dark forces that'll pursue them. After a near-deadly battle, the hero returns to their community with learnings and 'boons'.

Thirty years of study led Christopher Booker to assert the existence of seven recurring plots in story. He calls them: Overcoming the Monster; Rags to Riches; The Quest; Voyage and Return; Rebirth; Comedy; and Tragedy. Each plot, he argues, tends towards five structural movements: the call to action, a dream stage in which everything goes well, a frustration stage, a descent into nightmarish conflict and finally a resolution. Following Jung, Booker outlines a character transformation he believes ubiquitous. At the story's start the protagonist's personality will be 'out of balance'. They'll be too strong or weak in the archetypal masculine traits of strength and order, or the archetypal feminine traits of feeling and understanding. In the happy resolution of the third act, the hero achieves 'the perfect balance' of all four traits and finally becomes whole.

In his fascinating book on story structure *Into the Woods*, John Yorke argues for the existence of an essential 'midpoint' in story. Partly inspired by Gustav Freytag's nineteenth-century analysis of Ancient Greek and Shakespearean drama, Yorke argues that an event occurs almost halfway through 'any successful story' during which something 'profoundly significant' takes place that transforms the story and its protagonist in some irreversible way. King Lear's scene on the stormy heath, when he raged and despaired over his sudden realisation of what his evil daughters had wrought, is a classic midpoint. Yorke additionally believes there's a hidden symmetry in story, in which protagonists and antagonists function as opposites with their rising and falling fortunes mirroring one another.

The Hollywood animation studio Pixar is home to some of the most successful mass-market storytellers of our age. 'Story artist' Austin Madison, who's worked on blockbusters including *Ratatouille*, *Wall-E* and *Up*, has shared a structure he says all Pixar films must adhere to. The action starts with a protagonist who has a goal, living in a settled world. Then a challenge comes that forces them into a cause-and-effect sequence of events that eventually builds to a climax that demonstrates the triumph of good over evil and the revelation of the story's moral.

The arrival of 'big data' has led to a new era of story analysis. Researchers downloaded 1,327 of the most popular works of fiction available via Project Gutenberg, an online platform that makes available out-of-copyright novels. With the use of an algorithm, they sliced the books into 10,000-word sections and measured the emotional temperament of the language contained in each. They found stories tended towards six 'emotional arcs': Rags to Riches (characterised by rising emotion); Riches to Rags (tragedies, characterised by falling emotion); Man in a Hole (a fall then a rise); Icarus (a rise then a fall); Oedipus (fall, rise, fall). The most commercially successful emotional arcs, they found, were Icarus, Oedipus and 'two sequential Man in a Holes'.

Another big-data analysis was carried out by publishing executive Jodie Archer and Matthew Jockers of Stanford University's Literary Lab. After their algorithm had been to work on 20,000 novels it could predict a *New York Times* bestseller with an accuracy of eighty per cent. Fascinatingly, the data supported the life's work of Christopher Booker, whose seven basic plots did, indeed, emerge. What also emerged was an indication of what people are most curious to read about. The 'most frequently occurring and important theme' of bestsellers was 'human closeness and human connection', an apposite interest for a hyper-social species.

Archer and Jockers were especially interested in the novel *Fifty Shades of Grey* by E. L. James, whose 125-million-selling success baffled many in the publishing industry. Some assumed it was successful because of its BDSM subject matter, but a textual analysis revealed that sex wasn't actually its dominant theme. 'The novel is not so much outright erotica, but is instead a spicy romance that has the emotional connection between its hero and heroine as its central interest,' they wrote. What actually drove the action was 'the constantly recurring question of whether or not Ana will submit'. The plot was powered, as all plots should be, by the dramatic question: who was Ana going to be?

When Archer and Jockers laid out the plot of *Fifty Shades of Grey* on a graph, it turned out to take an intriguing and rare form. It made a roughly symmetrical pattern that travelled across five peaks and four valleys, each of which came regularly. Unusual as it was, it was strikingly similar to another novel that seemed to come from nowhere and into sales of dozens of millions: Dan Brown's *The Da Vinci Code*. 'The distance between each peak is about the same, and the distance between each valley is about the same, and finally, the distances between peaks and valleys are about the same,' they wrote. 'Both novels have mastered the page-turner beat.'

All these plot designs embrace the three-act shape of crisis, struggle, resolution. When unexpected change happens to the right person, it ignites a drama that eventually comes to a conclusion. What these theorists often disagree about is the events of act two. But I suspect that none of these plot designs is actually the 'right' one. Beyond the basic three acts of Western storytelling, the only plot fundamental is that there must be regular change, much of which should preferably be driven by the protagonist, who changes along with it. It's change that obsesses brains. The challenge that storytellers face is creating a plot that has enough unexpected change to hold attention over the course of an entire novel or film. This isn't easy. For me, these different plot designs represent different methods of solving that complex problem. Each one is a unique recipe for a plot that moves relentlessly forwards, builds in intrigue and tension and never stops changing.

These recipes work. But the problem with recipes is that, every time you follow one, you get the same bloody cake. Perhaps a more creatively freeing way of looking at plot is as a symphony of change. There's the top level of cause-and-effect in which all the obvious action and drama plays out. There's the second level in which characters are altered in surprising and meaningful ways. As well as this, the characters' understanding of their situation can change. The characters' plan for achieving their goal can change. The character's goal can change. A character's understanding of themselves can change. A character's understanding of their relationships can change. The reader's understanding of who the character is can change. The reader's understanding of what's actually happening in the drama can change. The secondary major (and even tertiary) characters can change. Information gaps can be opened and teased and closed. And so on. An efficient and immersive plot is one in which change is constant and taking place on many layers in harmony, with every new movement pushing the intertwined characters relentlessly towards their conclusions.

Which forms of change are deployed, and when, is a creative decision that depends partly on the kind of story that's being told. Police-procedural drama, for example, depends heavily on changes in the reader's understanding of what's really happening, which tend to dance exhilaratingly around what the Detective Inspector knows. This is change that plays with information gaps – curiosity is aroused and toyed with throughout act two and then finally satisfied. Much of the change in *The Remains of the Day*, meanwhile, takes the form of the reader's understanding of Stevens, a character to whom nuance and colours (many of them dark) are progressively added, often with the use of flashbacks.

If this second form of change is more profound and memorable it's because it more directly connects with that elemental dramatic question. Who is Stevens? Who's he going to be? The answer doesn't stop changing until Ishiguro's very final page.

4.2

The job of the plot is to keep asking the dramatic question. It does this by repeatedly challenging and gradually breaking the protagonist's model of who they are and how the world works. This requires pressure. These models are tough. They run to the core of the character's identity. If they're going to crack, the protagonist needs to hurl themselves at the drama. It's only by being active, and having the courage to take on the external world with all its challenges and provocations, that these core mechanisms can ever be broken down and rebuilt. For the neuroscientist Professor Beau Lotto it's 'not just important to be active, it is neurologically necessary'. It's the only way we grow.

When the data scientist David Robinson analysed an enormous tranche of 112,000 plots including books, movies, television episodes and video games, his algorithm found one common story shape. Robinson described this as, 'Things get worse and worse until, at the last minute, they get better.' The pattern he detected reveals that many stories have a point, just prior to their resolution, in which the hero endures some deeply significant test. For one final, decisive time, they're posed the dramatic question. It's the crucial moment in which they have the chance to become someone new.

In archetypal storytelling, especially as it emerges in fairy-tales, myths and Hollywood movies, this event often takes the form of some life-or-death challenge or fight in which the protagonist comes face-to-face with all that they most dread. This is symbolic of what's taking place in the second, subconscious layer of the story. Because the events of the drama are specifically designed to strike at the core of their identity, the thing they need to change is precisely that which is hardest, and that they least want to confront. The flawed models they're required to shatter run so deep that it takes an act of almost supernatural strength and courage to change them.

The psychologist and story theorist Professor Jordan Peterson talks of the mythic trope in which a hero makes final battle with a dragon that's hoarding treasure. 'You confront it in order to get what it has to offer you. The probability is that's going to be intensely dangerous and push you right to the limit. But you don't get the gold without the dragon. That's a very, very strange idea. But it seems to be accurate.'

That gold is your the reward for accepting the fight of your life. But you only get it if you answer story's dramatic question correctly: 'I'm going to be someone better.'

4.3

How does a story end? If all story is change then it naturally follows that a story ends when the change finally stops. From the ignition point onwards, the protagonist has been in a battle to reimpose control over the external world. If the story has a happy ending the process will be successful. Their brain's model of the external world, and its theory of control, will have been updated and improved. They'll finally be able to tame the chaos.

Control, as we have already discovered, is the ultimate mission of the brain. Our hero-making cognition always wants to make us feel as if we have more of it than we actually do. When study participants were faced with a machine that issued rewards at random, they concocted elaborate rituals with its levers, convinced they were able to control when it paid out. Another test found that participants given electric shocks could withstand more pain simply by being told they could stop it at will. Random and uncontrollable shocks, meanwhile, led to psychological and physiological decline.

To lose our sense of control is to suffer the loss of the sense of ourselves as an active heroic character, and this leads to anxiety and depression and worse. Desperate to avoid this, the brain spins its compelling, guileful and simplistic story of heroic us. 'A critical element to our well-being is how well we understand what happens to us and why,' writes psychologist Professor Timothy Wilson. Happy people have reassuring narratives of self that account for why bad things have happened to them and which offer hope for the future. Those who 'feel in control of their lives, have goals of their own choosing and make progress towards those goals are happier than people who do not'.

Brains love control. It's their heaven. They're constantly

battling to get there. It's surely no coincidence that control is the defining quality in the hero of the world's most successful story. The star of the majority of religious sagas is 'God'. He can do anything. He always knows what to do, He knows what's coming, He knows what's happened and He has unrestricted access to everyone's most private gossip.

Our craving for control explains why the endings of archetypal stories are so deeply satisfying. In tragedies such as *Lolita*, the protagonist answers the dramatic question by deciding not to become someone better. Rather than discovering and fixing their flaws they embrace them yet further. This causes them to enter a catastrophic spiral of model-defending behaviour that loosens their control over the external world more and more, leading to inevitable humiliation, ostracisation or death. Such an ending transmits the profoundly comforting signal, to the reader, that divine justice truly exists and is inescapable, and that there's control in the chaos after all.

Stories such as Lars von Trier's *Dancer in the Dark* take advantage of our wired-in lust for control by deliberately and cruelly not satisfying it. When her money is stolen by the selfish policeman, the selfless immigrant Selma Ježková's attempts at regaining control over the external world cause her to spin yet further into disarray. The plot ends with her death by hanging in a prison. This is not what we want. In refusing to fulfil our tribal desire for justice and restored control, Von Trier leaves his audience in a state of devastation. By doing so, he successfully and powerfully makes his political comment on the treatment of the vulnerable by the United States.

The ending of Damien Chazelle's screenplay *La La Land* both satisfies and subverts our need for control. His romantic comedy follows two protagonists, one of whom is desperate to become a famous actress, the other a lauded jazz musician. When the plot poses each of them the dramatic question, they ultimately choose their ambitions over each other. In the wonderfully effective ending we're happy to discover their dreams came true and yet sad they lost each other in the process. The ending works because the dramatic question is answered decisively and it feels true to who the characters are, and yet the viewer is left drowning in lovely, longing bitter-sweetness. They achieved control and lost it too.

The butler Stevens's story ends by promising us, subtly but surely, that his ability to control reality will transform. Extended flashback sequences in Ishiguro's *The Remains of the Day* show us the melancholy consequences of his loyalty not only to the value of dignity in emotional restraint, but also to his former employer, Lord Darlington, who emerges as an anti-Semite and Nazi appeaser. Events on Stevens's road trip to Cornwall, where he's to meet with his former housekeeper Miss Kenton, cause various knocks to his internal model of the world, but he remains stubbornly true to it.

When he finally meets Miss Kenton she admits she was once in love with him. On hearing her confession, Stevens admits to the reader that his 'heart was breaking'. He nevertheless fails to share his feelings with Kenton herself, even as her eyes brim with tears. His model of the world, and its theory of control, has it that to show anything but dignity in emotional restraint is to invite chaos. He simply cannot do it.

The story's closing paragraphs take him to Weymouth pier where crowds have gathered in what remains of the day to see the electric lights turned on. Finally, Stevens concedes he was wrong about Lord Darlington who, he admits, made 'mistakes'. He reflects that his position of servitude demanded loyalty to whatever view of the world Darlington chose. 'What dignity is there in that?' he asks.

Moments later, he's surprised to realise the people chatting behind him are not friends or family members but strangers gathered to watch the lights. 'It is curious how people can build such warmth among themselves,' he says. Wondering how it happens, he concludes it's likely down to the 'bantering skill' that his new American employer enjoyed so much but which he'd given up trying to master. 'Perhaps it is indeed time I began to look at this whole matter of bantering more enthusiastically,' he says. 'After all, when one thinks about it, it is not such a foolish thing to indulge in – particularly if it is the case that in bantering lies the key to human warmth.'

In the book's final page Stevens makes a commitment to change that might be trivial to anyone else but for him means wrestling dragons. His internal model of the world has been recognised as wrong and the reader is left in the lovely glow of the implication that his ability to control the external world will be improved and, as a result, he'll receive the golden treasure of transformation. The ending of his story is a happy one.

An archetypal happy ending can be found in the closing paragraphs of Ken Kesey's *One Flew Over the Cuckoo's Nest*. Set in a psychiatric institution in 1950s, the novel is narrated by the native American patient Chief Bromden, whose model of the world is, like Mr B's, pathologically delusional.

When we meet him, he believes reality itself is controlled by a strange hidden mechanism he calls the Combine. His theory of control has it that he has no control at all. Bromden doesn't talk, he just sweeps repetitively in the corner and listens. His model of the world is challenged and rebuilt by the arrival of the charismatic and rebellious McMurphy, who ends up being cruelly lobotomised. In an exceptionally moving ending, Bromden mercifully euthanises the friend who helped him heal. He then tears a heavy control panel out of the ground, hurls it through a window and leaps into the moonlit sky, leaving us with the words, 'I been away a long time.'

Back at the story's start, Bromden appeared to be in hospital again, perhaps caught as an AWOL or having fallen ill once more. But the story ends where it does because that's the blissful, fleeting instant in time in which Bromden has complete control over both levels of story: over the external world of the drama and the internal world of who he is. For one blissful, perfect moment, he has control over everything. He has become God.

The perfect archetypal ending takes the form of 'the God moment' because it reassures us that, despite all the chaos and sadness and struggle that fills our lives, there is control. There's no more reassuring message than this for the story-telling brain. Having been picked up in act one, and hurled around the drama, we're put back down again in the best possible place. The psychologist Professor Roy Baumeister writes that 'life is change that yearns for stability'. Story is a form of play that allows us to feel we've lost control without actually placing us in danger. It's a rollercoaster, but not one made from ramps, rails and steel wheels, but from love, hope, dread, curiosity, status play, unexpected change and moral outrage. Story is a thrill-ride of control.

4.4

To live in a hallucination trapped inside a skull is to feel, in the words of neuroscientist Professor Chris Frith, like 'the invisible actor at the centre of the world'. We're that single point of focus at which everything meets: sight, sound, smell, touch, taste, thought, memory and action. This is the illusion story weaves. Writers create a simulacrum of human consciousness. To read a page in a novel is to move naturally from visual observation to speech to thought to the recollection of a distant memory, back to visual observation again, and so on. It is, in other words, to experience the consciousness of the character as if we *were* them. This simulacrum of consciousness can become so compelling it nudges the reader's actual consciousness backwards. When we're lost in story, brain scans suggest the regions associated with our sense of self become inhibited.

As the story sends us on its thrilling rollercoaster of control, our bodies respond accordingly, experiencing its events: heart rate goes up, blood vessels dilate, changing activations of neurochemicals such as cortisol and oxytocin have powerful effects on our emotional states. We can become so replaced by the storyteller's simulated model-world that we miss our train stop or forget to go to sleep. Psychologists call this state 'transportation'.

Research suggests that, when we're transported, our beliefs, attitudes and intentions are vulnerable to being altered, in accordance with the mores of the story, and that these alterations can stick. 'Research has demonstrated that the transported "traveller" can return changed by the journey,' write the authors of a meta-analysis of 132 studies of narrative transportation. 'The transformation that narrative transportation achieves is persuasion of the story-receiver.'

In the 1960s, the novel *One Day in the Life of Ivan Denisovich* by Aleksandr Solzhenitsyn dragged its readers through the experiences of an ordinary prisoner in one of Stalin's gulag camps, shocking the Communist citizens of the Soviet Union. During the nineteenth century, slave narratives brought white readers into the lives of those trapped in bondage in the southern states of America. Books such as *The Narrative of the Life of Frederick Douglass* sold by the tens of thousands and gave abolitionists a mighty weapon, while Harriet Beecher Stowe's bestseller *Uncle Tom's Cabin* helped precipitate the American Civil War.

Transportation changes people, and then it changes the world.

4.5

We all inhabit foreign worlds. Each of us is ultimately alone in our black vault, wandering our singular neural realms, 'seeing' things differently, feeling different passions and hatreds and associations of memory as our attention grazes over them. We laugh at different things, are moved by different pieces of music and transported by different kinds of stories. All of us are in search of writers who somehow capture the distinct music made by the agonies in our heads.

If we prefer storytellers with similar backgrounds and lived experiences to our own, it's because what we often crave in art is the same connection with others that we seek in friendship and love. It's only natural if a woman prefers books by women or a working-class man prefers working-class voices: storytelling will always be full of associations that speak directly to particular perspectives.

Take this first sentence: 'The North Carolina Mutual Life Insurance agent promised to fly from Mercy to the other side of Lake Superior at three o'clock.' To this middle-aged Kentishman it's a fine enough opener, but has little resonance beyond its surface facts. But readers with a similar background to its author, Toni Morrison, might know the North Carolina Mutual Life Insurance agency was one of the largest African-American owned companies in the United States, and one founded by a former slave. Morrison also hoped the reader would pick up on a sense of movement from North Carolina to Lake Superior that, she writes, 'suggests a journey from South to North – a direction common for black immigration and in literature about it'.

But just because books by people like us can ring with greater personal meaning doesn't mean we should stay in our silos. It doesn't require a forbidding amount of historical or cultural knowledge to enjoy Morrison's *Song of Solomon*. Psychologists have examined the effects of storytelling on our perceptions of tribal 'others'. One study had a group of white Americans viewing a sitcom, *Little Mosque on the Prairie*, that represented Muslims as friendly and relatable. Compared to a control group (who watched *Friends*) they ended up with 'more positive attitudes towards Arabs' on various tests – changes that persisted when re-tested a month later.

Story, then, is both tribal propaganda and the cure for tribal propaganda. In Harper Lee's *To Kill a Mockingbird*, Atticus Finch advises his daughter Scout that she'll 'get along a lot better with all kinds of folks' if she learns a simple trick: 'You never really understand a person until you consider things from his point of view . . . until you climb into his skin and walk around in it.' This is precisely what story enables us to do. In this way, it creates empathy. There can hardly be a better medicine than that for the groupish hatred that comes so naturally and seductively to all humans.

And yet it's sometimes argued that a storyteller who climbs into the skin of a person of a different gender, race or sexuality is guilty of a kind of theft – that of appropriating and unjustly profiting from another's culture. Storytellers who attempt such feats of imagination have a heightened obligation towards truth, to be sure. But I don't believe they're the enemies of peace, justice and understanding. On the contrary, I fear it's those who rage against them who'll end up dividing us further. Smart people will always be able to construct persuasive moral arguments to defend their beliefs, but calls to keep strictly within the bounds of one's group seem to me to be little more than chimpish xenophobia.

Story should not respect such boundaries. If tribal

thinking is original sin, then story is prayer. At its best, it reminds us that, beneath our many differences, we remain beasts of one species.

4.6

The lesson of story is that we have no idea how wrong we are. Discovering the fragile parts of our neural models means listening for their cry. When we become irrationally emotional and defensive, we're often betraying the parts of us that require the most aggressive protection. This is the place in which our perception of the world is most warped and tender. Facing these flaws and fixing them will be the fight of our lives. To accept story's challenge and win is to be a hero.

4.7

The consolation of story is truth. The curse of belonging to a hyper-social species is that we're surrounded by people who are trying to control us. Because everyone we meet is attempting to get along and get ahead, we're subject to near-constant attempts at manipulation. Ours is an environment of soft lies and half smiles that seek to make us feel pleasant and render us pliable. In order to control what we think of them, people work hard to disguise their sins, failures and torments. Human sociality can be numbing. We can feel alienated without knowing why. It's only in story that the mask truly breaks. To enter the flawed mind of another is to be reassured that it's not only us.

It's not only us who are broken; it's not only us who are conflicted; it's not only us who are confused; it's not only us who have dark thoughts and bitter regrets and feel possessed, at times, by hateful selves. It's not only us who are scared. The magic of story is its ability to connect mind with mind in a manner that's unrivalled even by love. Story's gift is the hope that we might not be quite so alone, in that dark bone vault, after all.

o

THE SACRED FLAW APPROACH

This is a technique that's been developed principally in my writing classes since 2014. It's an attempt to incorporate essential Science of Storytelling principles into a practical, step-by-step method for creating effective and original stories.

I call it an 'approach' because it's a series of exercises to undertake principally as you approach the actual writing of your first draft. The essential idea is to build a protagonist, in all its fascinating weirdness and damage, in the same way that a brain builds a self. By going through a relatively straightforward series of steps, we can aim to discover a character who is interesting and original and has, surrounding them, a cast of compelling and necessary secondary characters. Incorporated into the process are some well-known and popular creative writing exercises which you might have encountered before.

While you're working with the Sacred Flaw Approach, it's important to remember a couple of things. First, I'm not suggesting in any sense that this is the *only way* to make story.

This is simply one route which the people who've attended my class have found useful. Second, it doesn't need to be followed religiously. The demands of your particular piece might make some parts of this framework irrelevant or inappropriate. You might reach a point where you no longer need it. It's really just a guide to help you think in the right direction. The only thing that matters is that it helps.

The method's focus is on character because, for me, this is where all storytellers should begin their serious creative endeavours. Character work is essential whether you're writing an art-house film or some plot-heavy genre fiction. If your passion is for the thriller or bonk-buster or action-adventure story, your plots clearly need to be tight and efficient and superior. But if you disregard character they risk becoming predictable. In life our 'plots' emerge out of who we are. It's the *active decisions we make* that create the events of our days. These decisions reflect our character – our values, flaws, personality and goals. It's in this way that the lives we lead emerge out of the people we are. This is true in life and it should also be true in story.

SACREDNESS

During my research into the storytelling brain, I was lucky enough to interview the famous psychologist Professor Jonathan Haidt. He told me something I've never forgotten: 'Follow the sacredness. Find out what people believe to be sacred, and when you look around there you will find rampant irrationality.'

Rampant irrationality! This is exactly what we, as story-tellers, should be hunting in our characters. In order for them to change, our protagonists need to start off broken. When we meet them, in act one, they should be immersed in a reality of rampant irrationality without really being aware of it. This is not to say they should believe the earth is a painted cauli-flower or there's a vampire camping in their sock-drawer. They're not *crazy* crazy. They're ordinarily crazy – a person you might meet in your everyday life who's become locked into some belief or behaviour that's somehow damaging them, even if they can't see it.

In order to locate the thing they're irrational about, we need to ask what they make sacred. The things we make sacred are, to a great extent, the things that come to define us. This, I believe, is the secret of unlocking the truth of a character. When other people think of us – when they're asked what we're like – our sacred belief will probably be the first thing that pops into their minds. It will be how they describe us to a stranger. And because sacredness is the source of 'rampant irrationality', it's also probably going to be the cause of misery, error and distress. This is the stuff of story. It's exactly what we're hunting.

A fictional character's 'sacred flaw', then, is the broken part of them that they've made sacred. In *The Remains of the Day*, the butler James Stevens had made the idea of English dignity in emotional restraint sacred. This is where we meet him in act one. Early in *Citizen Kane*, we watch Charles Foster Kane make the idea of himself as a selfless warrior for the 'common man' sacred – a faulty belief that powered the rest of his journey. Likewise, the early sequences of *Lawrence of Arabia* portray T. E. Lawrence making the idea that he is an 'extraordinary' man sacred – and then we're dragged unforgettably through the consequences of this irrational belief.

These were faulty concepts that became built into these characters' neural models of reality. They struggled to see past them. They defined who they were. The point of the plots was to test and retest and ultimately break these sacred ideas apart. That's what the drama was for. That's what made those stories gripping.

The cases of Citizen Kane and Lawrence of Arabia demonstrate that the sacred flaw doesn't necessarily have to be fully in place at the story's start. But don't forget those protagonists were still damaged when we met them – Kane with his childhood trauma and Lawrence with his cocky, rebellious vanity. Their personalities were such that, when the events of the plot started flying at them, it became inevitable that these flaws rapidly developed and came to possess them. Additionally, these were tragic plots, so the development of the flaw, rather than the process of it being healed, took centre stage in their stories. If we're aiming for a happier ending we're more likely to meet someone more overtly flawed on page one, and then root for them as they slowly work out who to become in order to heal.

For the purposes of this exercise we're going to work towards this model – a character who's in the grip of their sacred flaw on page one, as this tends to make for the more dramatic beginning. There's nothing stopping you starting your story earlier, however, especially if you're going for a tragic theme. But keep in mind that your character still needs to be broken in a specific and purposeful way at the moment we meet them. Their flaw needs the potential to develop into something damagingly irrational.

Our job, then, is to find the sacred flaw in our protagonist. Once we've found it, we need to build a life and a self around this person. We need to work out what effect this flaw has had on them. What ramifications has it had for their family life, their romantic life and their working life? What benefits has adherence to this flaw brought them? And what costs? By doing so, an entire world can be conjured around this tiny flaw. This will be the world of our story.

EMBRACE THE REWIND

With every class I teach, there are usually one or two writers who politely resist this process. When I work with them, I sometimes sense the problem is that they've rather fallen in love with their protagonists. They've lived with them for months and maybe years of drafting and redrafting and they don't want to closely define them because they're *this* and they're also *this* and they're *this* and *this* and *this* and, oh my God, they're just *amazing*! The last thing they want is to assign to them any flaws.

For some of these students, I suspect what's secretly holding them back is that the protagonist actually is them. The more work they do on that character, the further that character moves away from who they are. As strange as it might sound, this process can cause them some emotional pain, almost as if they're losing a loved one. But it's pain they must endure. Unless it's overcome, this problem can be fatal to their creativity. A storyteller needs spine. They have to make hard and clear decisions about their characters, even if those decisions are left somewhat ambiguous on the page. Underpinning every gripping scene in their story is that fundamental dramatic question: who is this character really? If the author doesn't know, the reader is likely to sense it and grow confused, frustrated and uninterested.

A further issue is that starting out with little more than a close definition of a single flaw can feel reductive. In the early stages of the process, it's almost as if we're sketching cartoon characters. But starting with absolute specificity is incredibly useful. Over and over and over again I've found this somewhat paradoxical truth: the more tightly we define the sacred flaw, the more complex and unique the character that explodes out of it.

A final, practical problem is that storytellers resist focussing on character because the source of their inspiration, and their excitement about their project, is not a character at all. There are three main routes into a story idea that don't come from character – a milieu, a what-if and an argument.

THE MILIEU

Here's a reasonable milieu – scientists have found the cure for death and the earth is overflowing with humans. It feels like it could be the basis of some big-budget TV series. But the problem is, it's not a story, it's a *setting* for a story. The risk is that the writer feels most of their creative heavy lifting is now done and, having thought of a dark and compelling milieu, they just have to fill it with some thrilling action. So here's the haggard cop and here's the ballsy sex worker and here's the brave but beleaguered politician and here's some cool CGI panning shots of a foggy night in the rammed metropolis.

None of this is good. To move beyond cliché requires specificity. What specific part of this deathless world can we zoom in on? Well, what's happening to the earth's resources? Is this a place of extremely heightened inequality in which only the rich can afford to eat fresh food and see the ocean? That could be an interesting line to pursue. Or perhaps we could think about the people who decide that, despite the cure, they actually *want* to die. There'd presumably be a booming euthanasia industry. There'd be peripheral industries too – what if there was a paradise island where the tired-of-life could go to, in their last week, in order to live out their wildest dreams? What kind of weird human dramas could happen in a place like that? Perhaps the story could be about intergenerational war, as 200-year-old humans with 200-year-old political views fight the new progressive generation?

Or what if our story followed a renegade scientist who wanted to save the planet from this runaway plague of humans? And this person is attempting to destroy the cure for death? This could make for an interesting subversion, in which the plucky selfless hero is the person who's trying to kill everyone. She'd surely be likely to suffer from some massive internal conflict over her project.

We're getting closer. Let's go with the scientist. I can picture her immediately. She's a beautiful, gutsy, empowered biologist who lives alone and like a drink and struggles against the establishment. Are you bored yet? We're *still* in the land of cliché. The only way to escape it is to work out precisely who this person is, how she's damaged and therefore what specific battle the plot must create for her.

THE WHAT IF?

What if a world-famous celebrity became their own look-alike? He decided, for whatever reason, to escape Hollywood and hide out in a small regional town. (Maybe there's been scandal? Maybe he inherited an apartment in this obscure country town from an aunt and it was the only place nobody would look for him?) None of the townspeople would expect to see *him* in this place. On his first day he bumps into the owner of a beleaguered lookalike agency who realises he 'kind of' looks like the actor he actually is. He talks the celebrity into doing a last-minute job at a party that evening. He'll be serving free tequila shots to a hen party.

This is a reasonable 'what if' for perhaps a black or a broad comedy. I can picture the protagonist immediately. He's past his peak but still handsome, sarcastic, dry, but lovable some-where deep down. On his first gig he's horrified to discover how much he's hated by the public. What he needs, in order to heal, is to reconnect with authentic people. Back in Hollywood he's just become engaged to a spoiled, skinny starlet. But then in walks kooky barmaid Serena. She drives a beaten-up old Mini. Some of her hair is pink. Are you bored yet? Once again, we're drowning in cliché. How else is this 'what if' going to become a story that moves us and surprises us and feels as if it's saying something real, if not by digging right down into the unique character of the protagonist?

THE ARGUMENT

Increasingly, writers come to the course who are angry about something that's happening in the world. They want to write a story that highlights some perceived societal problem. Say you're angry about the US healthcare system, so you decide to write a kind of healthcare version of Oliver Stone's *Wall Street*. It centres on a Gordon Gekko type who ramps up the price of an essential medication. Fine. The risk is that, if you don't do the necessary character work, 'a healthcare version of Oliver Stone's *Wall Street*' is exactly what you're going to end up with.

No matter where your inspiration came from or what kind of story you want to write, I believe no harm can ever come from rewinding to character. All stories are ultimately about how people change and your story will be all the better for having a properly designed protagonist and principal characters, all of whom have plenty of potential to do just that.

THE IGNITION POINT (see section 2.5)

Our first task is to work up to the ignition point. This is that magical moment, in a story, when we suddenly become engaged. The reader sits up, in the narrative, tense and expectant and excited. It's triggered when just the right event happens to just the right character – when we sense that an unexpected change has taken place that, no matter how slight it is, strikes to the core of that character's deepest flaw. This event triggers them. It makes them react in a surprising and specific way which makes us realise something unusual is afoot. It makes them act. It launches them into the plot.

If you're starting completely afresh, take a blank piece of paper and write a list of unexpected changes that might occur. It could be an ordinary everyday event or it could be magical or extraordinary. Start wherever your mind wants to go. Once you've finished, go back down your list and, next to each event, write a brief description of an ideal person for that change to happen to. It should be the first ripple in a flood which ultimately has the power to completely overturn who that person is. The important question is, why are we making this event happen to *this* person? Why will this end up being the start of a chain of events that will break who they are?

If you're starting with a milieu, a what-if or an argument, rewind back to a specific character and pair them with a specific event. Roughly who are they? What's their sacred flaw? Let's consider how this might work for an argument. One example of an argument you could build a story around is: war makes monsters of men. This is *Lawrence of Arabia*. Who is the character whom war and violence would most trigger? That is, who would be most likely to be psychologically overturned by war? It might be someone who has narcissistic tendencies and gets carried away with themselves.

They would also be rebellious and not enjoy following orders. This is the film's protagonist, T.E. Lawrence. He was uniquely vulnerable to the situation in which he found himself. When the drama and the character met they created a spark. When he first encountered violent conflict he reacted in an unusual way, by scolding and insulting the killer. This curious reaction signalled to us that something interesting was afoot. That's the ignition point.

If you're working on a list, there should hopefully be one combination of change and character that excites you more than the others. It might give you a tingle of promise and fascination. You could even feel vague flashes of scenes pop into your head. Choose that one.

Don't worry about getting it right the first time. This won't happen. From here on in, you'll hopefully enter a process of continual revision and refinement. You have a very rough outline of a protagonist. You have a very rough idea of the unexpected change that's going to make the ignition point. These should keep evolving as you draw further into your character work. The better you come to know your character, the more specific you can make the change event. The more specific your change event becomes, the more you can refine the character to match it, and so on.

ORIGIN DAMAGE (see section 3.11)

Now we know roughly how our story will begin. But we're still dealing in outlines. It's time to get specific and work out who our protagonist is. The flaws that characterise us typically have their origins in events that happen in the first two decades of life. This is when the brain is in its heightened state of plasticity and its neural models of the world are still being formed. Because these traumatic experiences get built into the structure of our brains, they become folded into who we are. We internalise them. They become part of us.

It's common in story for there to be a moment when the protagonist reveals their origin damage to another character, or we see it in flashback, and we gain a sudden insight into them. Although it's by no means compulsory that such scenes be included, I believe it's important that the writer knows what these moments are. This task involves working out exactly when the damage that created their flaw took place. Write out the scene in which your character came upon this faulty idea. And do it in full – the characters, the setting, the dialogue.

By fully imagining this moment of damage, your character will begin to come alive in your mind. You can almost hear them taking their first tentative breath. It's now that you start to get beyond cliché. Fully conjuring the scene forces you into thinking specifically, so it's not merely a case of 'her father beat her' or 'his mother didn't love him'. This is an actual, detailed incident that had a highly specific outcome.

It doesn't necessarily have to be obviously traumatic. For James Stevens in *The Remains of the Day*, it was hearing an inspirational story about his father behaving with supernatural levels of emotional restraint. For Amy Dunne in *Gone Girl*, it was her parents' popular 'Amazing Amy' children books that made her believe, 'People only love me when I appear perfect and amazing' (although we're not given a specific incident in the novel). But it could, just as easily, be a moment of pain. As we've discovered, because of our tribal evolution, experiences of being ostracised and humiliated are tremendously hurtful for humans. Perhaps the origin of their damage lies in a moment in which such feelings were powerfully felt?

Now that you have the scene, you can nail the flaw down with greater precision. Exactly what did this incident make them believe? What damaging idea did it create? There are various ways you can conceptualise it but, essentially, their sacred flaw should be a mistake about themselves and how the human world works. You might think of it as a statement that begins in one of the following ways:

- *I'm only safe when I . . .*

- *People only love me when I . . .*

- *The one thing nobody is ever allowed to know about me is . . .*

- *The most important thing of all in life is that I . . .*

- *The terrible thing about other people is . . .*

Or it could be some variation on these. The crucial point is that it should be a specific mistake that has serious future consequences for their interactions with other people. This sacred belief is the seed around which their flawed character grows. It ultimately explains why they're so uniquely triggered by the moment of unexpected change that sets off the story.

PERSONALITY (see section 2.1)

At this stage, you might also want to consider their personality type. What version of self do they become when you run them and their flaw through the filter of one of the 'big five' traits?

CONFIRMATION BIAS (see section 2.5)

Now we need to begin turning this flaw into a person and a life. This means allowing the character to internalise it in such a way that they don't see it as a flaw at all. We're going to mimic the process by which a brain does this. We have our moment of origin damage and the belief about the world that it created. In this next step, the character needs to see powerful evidence that their belief is correct. In life, we have a bias that means we only tend to notice things in our environment that confirm what we already instinctively believe. We're going to imagine such a moment of 'confirmation bias' now.

This means writing a detailed scene in which confirmation bias is in full effect. Something happens to the character in which they become fully convinced that this flawed belief is not flawed at all, but true. This should be a pivotal moment in their young life. It's also an opportunity to aim them towards the ignition point we've already outlined. You first came to think of them as the person they were at that moment of ignition – in a particular place and time, perhaps with a particular job. When sketching out these key formative moments of change, you should be setting them on the road to getting there.

This should be a scene that involves some jeopardy. Something's got to be at stake. They also should be active in it. They need to let this flawed belief guide their behaviour at a moment at which they're strongly challenged – and find it works out for them. This key incident makes them feel (or, at least, they're able to thoroughly convince themselves) that this flawed belief is correct. Not only is it correct, it's the *most* correct belief they can possibly imagine anyone ever having. As far as they're concerned it's the key to how they're going to behave, from now on and forever.

From this moment forwards, their flawed belief becomes sacred. It becomes how they see themselves in the context of the human realm. It becomes their key to controlling the world.

THE HERO MAKER (see sections 2.6 and 2.7)

The brain is a hero-maker. No matter how wrong we are, it excels at seducing us into believing that we're right. The next stage of the approach is about firming up the character and properly embedding their flaw into their neural models. Place your character in a position in which they're being challenged by authority over a decision they've made on the basis of their flaw – and that has got them into trouble. It could be a police officer. It could be a teacher. It could be a romantic partner (as long as they're sufficiently aggressive in arguing back). Really put them in a position in which they have to powerfully defend themselves and, by extension, their sacred flaw. Remember, the brain makes us feel heroic in various ways:

- *It makes us feel morally virtuous*

- *It makes us feel like a relatively low-status David being threatened by more powerful Goliaths*

- *It makes us believe we're deserving of more status*

- *It makes us believe we're selfless, somehow, and that our enemies are selfish*

Once again, use this confrontation as an opportunity to move your character towards the ignition point – that magical moment when the event takes place which triggers them into action. Make this incident a key part of their back story. It's a formative experience that partly explains who they are when we first meet them in act one.

As you did previously, it's important to fully imagine the scene. Write it out in detail. Your job is to inhabit your character, and their flaw, in such a way that you're arguing so well in the defence of the stupid decision they've made that you practically convince yourself. We're seeing their sacred flaw taking over who they are – controlling their decisions and behaviour. It's become a core part of their identity, one they'll fight to defend.

POINT OF VIEW (see section 2.3)

Try rewriting the James Baldwin passage on page 77, but from the point of view of your character. They're walking into a jazz club in Harlem in the 1950s. How do they experience it? What details do they fix upon in their environment? What's the hero-maker narrative in their head? Make them feel intimidated or threatened in some way. Perhaps someone directly challenges them. How do they talk to themselves about these feelings? How do they make themselves feel better? What do they do?

This is the person we're going to meet on the first page of your story. If you want to immediately persuade an editor or producer that you're an exceptional talent, I believe the best way of doing so is by hurling a fully imagined character right at them.

THEORY OF CONTROL (see section 2.0)

By now, I hope you'll know your character well enough that you have a real feeling for their theory of control. This is their brain's overarching strategy for getting what they want out of the human world. It's the totality of their flaw, their personality and life experience, which you've sketched out in your previous scenes. It's how they typically respond to all the daily challenges that life, and other people, present.

Now it's time to take them to your ignition point – that electric moment when your story begins. Your character's theory of control, and the sacred flaw it's built around, will have made a particular life for them. It'll have led them on a particular journey – into a particular job with a particular romantic history, into a particular neighbourhood and particular home with a particular front door with a particular colour and state of repair. They'll have particular values and particular friends and enemies. We'll need to imagine what they are.

Say we're making a three-hour biopic of the children's book *Mr Nosey*. His sacred belief is something like: 'I'm only safe if I know everyone else's business.' What career might this flaw have made for him? Perhaps he's a domestic cleaner for the rich and famous. Perhaps he's a social worker. Perhaps he's responsible for judging prospective foster parents and, because of his origin damage, he takes his role of peering into their deepest crannies all too seriously, which gets him into trouble.

The important thing to remember, at this stage, is that their sacred flaw (as far as they're concerned) will probably have largely been of benefit to them. So think about:

- *How do they get status from this flaw? How does it make them feel superior? How has it brought them professional standing? As odd as it might sound, even if your character has extremely low status and is even self-loathing, there will be a way in which their flaw makes them feel somehow better than other people. (If they simply think they're worthless, and wrong about all their most precious beliefs, they're probably not a vastly interesting character.)*

- *How has their sacred flaw brought them closeness with friends, colleagues or lovers? What do they think of this person's flaw? How do they enable it? How do they challenge it? How do they cope with it?*

- *What joy does it bring them? For example, when the bourgeois, status-obsessed Emma Bovary attends an opulent ball, she takes great pleasure at marvelling at all the symbols of status, such as their wealthy guests' complexions that are the kind that 'comes with money' and 'look well against the whiteness of porcelain'.*

But, although the character will be probably be in denial of it, this flaw will also represent a huge vulnerability for them. It will be damaging their life in some way, even if they can't see it.

- *What do they dread will happen if they act against their flaw? What, in their minds, will they lose, materially and socially?*

- *What will actually happen?*

- *What enemies has it made them?*

- *What hidden risks has their flaw caused them to develop, such as for their marriage or financial security?*

How far you allow your character to inhabit their flaw is your own creative decision. If it's taken too far, you risk ending up not with character but caricature. However, it's worth bearing in mind that many of the most memorable and popular characters in film and literature – the ones that seem to burst, Scrooge-like, from the screen or page utterly alive and compelling – are the ones who seem the most possessed by their mistaken idea.

All story is change, and the most important change of all that takes place is to your protagonist. The further you pull back the bow, at this stage, the further your narrative arrow will be able to fly. The more concentrated they are, the more tantalised we'll be on meeting them, because we'll sense that huge dramatic change must be coming.

This is the point at which you need to really let your imagination run wild. Try to keep a playful frame of mind. Take your time. Don't put too much pressure on yourself. Don't sit in front of a blank sheet of paper or an empty screen and expect genius to fill it up. Ideas are more likely to come when you're filling the dishwasher or walking the dog. Your job is to build a life for your character that's full of detail and potential jeopardy. It's to lead them up to that moment of change that ignites the story.

THE STORY IGNITES

Now let's go back to that ignition point. Are you still happy with it? Are the tribal emotions (see section 3.6) it'll trigger such that we'll feel empathetic for your character? That is, do they seem relatively selfless and low in status? Are there more powerful Goliaths ranged against them? Remember, these aren't compulsory tick-boxes – this is about balance. A character who is completely selfish, high in status and all-powerful is going to be hard to care about.

Is your moment of change still triggering enough for your protagonist? Do you want to adjust it, in light of how much better you know them? Remember, we're looking for a moment of unexpected change that connects with their flawed theory of control. It should hit them directly where they're most weird and vulnerable.

If it does, it will cause them to respond in an unusual way – it might be an extreme way or a surprising way or just an odd overreaction. It's because of this unusual response that reader and viewer will sense that something alchemical has happened and they've entered the realm of story.

The protagonist's unusual response will take the form of an active behaviour. They'll decide to deal with the ramifications of that change in a practical way. Because it's triggered their flaw, their attempt will not just fail but will bring them even more chaos. (If you decide that they are in temporary denial of what's happened to them, this denial should also somehow bring increased threat, or trigger another event that finally persuades them to act.) This is the first domino-fall in your character-driven plot. From thereon in, you're trying to build a series of episodes, some of which further your character's ambitions and some of which hinder them, but all of which are triggered by your protagonist's changing attempts to control the world.

As you move through the story, remember that every dramatic scene will likely pose your character the dramatic question: who am I going to be? (Or, from the reader's perspective: who is this person?). This is what your story's really about. The drama is a continual test for the protagonist. Are they going to be the old, flawed version of themselves? Or are they going to be someone new?

GOAL DIRECTION

It may well be that you're ready to go by now and head off into your story. If you're still unsure of your direction of travel, let's have another close look at the components of an ignition point.

Story operates on two levels. There are the dramatic changes that happen on the surface level, including all the physical action and dialogue. But then, beneath that, there are the changes that take place that involve the character's mind, especially their subconscious (see section 2.3). Because this process is focused on character, our attention thus far has primarily been on this second subconscious level. If you're still not clear where to go with your idea, there's a fair chance that it's the surface level of the drama that needs some more creative time thrown at it.

At the ignition point, something changes for a character that exposes their flaw. This creates a desire and that, in turn, creates an action. It's that action that drives the surface plot. So we need to locate a goal for your character to throw themselves at. This is the 'mission' they'll be on – it's what will likely be described in the blurb on the back of your book or form your screenplay's log-line.

- *The goal should be triggered by the ignition point. It makes them want something.*

- *The goal should be the product of their sacred flaw. What they decide they want has to come from the flawed core of their character. They react as they do, at the ignition point, because of who they are.*

- *The plot develops through their flawed attempts at trying to achieve this goal.*

Ask these questions:

- *What do they want most of all in the world? What do they imagine will make them happy forever?*

- *What do they subconsciously need? Of course, what they inevitably need is to identify their flaw and fix it. But what would that event actually look like? What form would it take on the surface, dramatic level of the plot?*

- *What must they do in order for that thing to happen?*

- *What will happen if your character does get what they most want in the world (but not what they need)? What unexpected problems will that bring? What will this teach them?*

You might also want to add some more drama and jeopardy into your ignition point. This will make your character behave even more irrationally and that will add drama. Try adding these complications to your scenario.

- *The event that's happened is the final straw. They've now snapped. Why?*

- *Today is the worst day imaginable for this to happen. Why?*

- *They have to act, in response to this event, immediately. Its ramifications have to be dealt with in the next 24 hours/seven days/month. Why?*

- *They can't allow anyone to know why this event is such an absolute disaster. Why?*

- *This event means they're going to have to take on the person they're most afraid of. Why?*

- *Something huge is at stake. What is it? Why is it so important to them?*

PLOTS

There's really no way to make proper character-driven plots other than to get stuck into the scenes and think through, beat by beat, all that might happen to your characters and how they might react. That said, your plot still needs an overall shape. It's at this stage you might find it useful to look at the various plot designs outlined in books on structure. I'd recommend *Into the Woods* by John Yorke, *The Seven Basic Plots* by Christopher Booker and *Story* by Robert McKee. Now you've done your character work, you won't allow these apparent magic keys to success to dictate where your story goes. Instead, they'll be in their proper place – waymarkers toward which you can choose, or not, to direct your self-propelling characters.

It could also help to have some definite endpoint to aim towards. You might find it useful to imagine the conclusion of your story, even if you accept that this might end up changing as you get to know your characters better and surprising things start happening to them. You know your protagonist's flaw. This means you can work out what it is, about your protagonist, that needs fixing.

- *Work out your rough character development. Divide it into four stages, each one moving forwards towards a healed version of who they are (assuming you're going for a happy ending. If you're writing a tragedy, they double down on their flaw, which ultimately leads to their comeuppance). What are these different versions of them? What must happen in the surface level of the plot to trigger them into being?*

- *Now throw one or two (temporary) reversals into the character development, as they revert back to negative behaviour and perhaps even start acting worse than they ever have before. Who or what would cause these to happen?*

- *Who's standing in their way? Who's helping them? Your protagonist will be changed by their encounters with these people – sometimes for the better, sometimes for the worse. Antagonists will often magnify the protagonist's sacred flaw while allies will open their eyes, a little, so they can see more clearly what it is they're getting wrong. Each encounter should somehow adjust who they are, in relation to their flaw. They should nudge their perspective about it, and send them off in a slightly new direction.*

- *Who can show them that those damaging things they experienced when they were growing up were wrong? How do they show them?*

The most important thing when writing is that your plot must be a symphony of change (see section 4.1). Change should be frequent and happen on more than one layer at once. Every truly dramatic scene will challenge your protagonist's flaw. The action will somehow serve to pose them that fundamental dramatic question, 'Who am I?' Are they going to be the old, flawed version of themselves? Or are they going to be someone new?

If your story has multiple protagonists you might find it useful to work through the Sacred Flaw Approach for each of them. I'd encourage you to consider how each protagonist connects with each other's flaw. They might have different versions of the same problem, which rub up against each other, making it better or worse, depending on the needs of the plot. In romantic comedies or buddy movies, the two protagonists often inhabit two opposing flaws. When they come together, they're healed.

THE GOD MOMENT (OR NOT)

Are you going to give your character a happy ending? That is, will your character work out how to heal their flaw? Or will it be a tragic ending? Will it be a bittersweet mixture of the two? If it's a happy ending, you'll be going for a 'God moment' (see section 4.3) in which your protagonist finally achieves what they need in the external world by finally mastering who they are in their internal world – and your character will have complete godlike control over everything for one blissful instant. If it's a tragic ending, they'll fail to heal their flaw and the consequences are likely to be grave and take the form of one of those tribal punishments – humiliation, ostracisation or death.

Of course, you might not want to do this and opt for an ambiguous, modernist ending. If so, it's *still* a good idea for you to have a firm grip on the duelling versions of who that character is, and to be skilful and deliberate in making your point, lest it feel as if you've simply opted out of making the decision through a lack of creative courage.

Whichever end you choose, if it's going to be satisfying (at least to a mainstream Western audience), it needs to deliver a firm answer to the dramatic question – we need to see, at the end of all the brilliant chaos and drama, who your protagonist *really* is.

GRIFFITH COLLEGE DUBLIN
SOUTH CIRCULAR ROAD DUBLIN 8.
Tel: 01 4150490 Fax: (01) 4549265
library@griffith.ie

A NOTE ON THE TEXT

This book is based on writing courses that were inspired by research that was undertaken for various writing projects. As such it incorporates portions of material, in mostly rewritten form, from my previous books *The Heretics* (Picador, 2013) and *Selfie* (Picador, 2017), as well as an essay that appears in the collection *Others* (Unbound, 2019).

This manuscript was proofread by two relevant experts, the neuroscientist Professor Sophie Scott and the psychologist Dr Stuart Ritchie. I'm extremely grateful to them both for their comments, corrections and assistance with fixing problems. Any errors that remain in the text are entirely my responsibility. If you've spotted any, I'd be grateful if you could inform me via my website willstorr.com so that I can investigate and, if necessary, correct any future versions of this book.

ACKNOWLEDGMENTS

They say 'everything is remix' and that's rarely been truer than in the case of this book. I'm incredibly grateful to all the story theorists and academics I quote in these pages, as well all the experts I've read over the years whose names I might have forgotten but whose insights have lingered.

Thanks, too, to my wonderful editor Tom Killingbeck and all at William Collins, my brilliant agent Will Francis, and the excellent Kris Doyle who edited the books that much of *The Science of Storytelling* is based upon. I'm massively appreciative, too, for the unerring support of Kirsty Buck at Guardian Masterclasses and Ian Ellard at the Faber Academy. My readers Professor Sophie Scott, Dr Stuart Ritchie and Amy Grier all gave invaluable advice – thank you for lending me your excellent brains.

My gratitude, too, to Craig Pearce, Charlie Campbell, Iain Lee, Charles Fernyhough, Tim Lott, Marcel Theroux, Luke Brown, Jason Manford, Andrew Hankinson, all at Kruger Cowne and, finally, my ever-patient and ever-beloved wife Farrah.

NOTES AND SOURCES

INTRODUCTION

Recent research suggests language evolved principally to swap 'social information': *Evolutionary Psychology*, Robin Dunbar, Louise Barrett, John Lycett (Oneworld, 2007) p. 133.

Some researchers believe grandparents came to perform a vital role in such tribes:
'Grandparents: The Storytellers Who Bind Us; Grandparents may be uniquely designed to pass on the great stories of human culture', Alison Gopnik, *Wall Street Journal*, 29 March 2018.

different kinds of stories: *The Origins of Creativity*, Edward O. Wilson (Liveright, 2017) pp. 22–24.

It is a 'story processor', writes the psychologist Professor Jonathan Haidt: *The Righteous Mind*, Jonathan Haidt (Allen Lane, 2012) p. 281.

Joseph Campbell's 'Monomyth': *The Hero with a Thousand Faces*, Joseph Campbell (Fontana, 1993).

I agree with the story analyst John Gardner who argues: *The Art of Fiction*, John Gardner (Vintage, 1993) p. 3.

CHAPTER ONE

1.1

'Almost all perception is based on the detection of change': Comment made by Professor Sophie Scott during review of manuscript, August 2018.

In a stable environment, the brain is relatively calm: *The Self Illusion*, Bruce Hood (Constable and Robinson, 2011) p. 125.

every one of them is as complex as a city: *Incognito*, David Eagleman (Canongate, 2011) p. 1.

speeds of up to 120 metres per second: *The Brain*, Michael O'Shea (Oxford University Press, 2005) p. 8.

150,000 to 180,000 kms of synaptic wiring: *The Domesticated Brain*, Bruce Hood (Pelican, 2014) p. 70.

John Yorke, has written: *Into the Woods*, John Yorke (Penguin, 2014) p. 270.

'There's no terror in the bang, only in the anticipation of it': *Halliwell's Filmgoer's Companion*, Leslie Halliwell (Granada, 1984) p. 307.

1.2

it's thought that we ask around 40,000 'explanatory' questions: *Curious*, Ian Leslie (Quercus, 2014) p. 56.

He writes of a test in which participants were confronted by a grid: 'The Psychology of Curiosity', George Lowenstein, *Psychological Bulletin*, 1994, Vol. 116. No 1. pp. 75–98.

There is a natural inclination to resolve information gaps: *An Information-Gap Theory of Feelings about Uncertainty*, Russell Golman and George Loewenstein (Jan 2016).

Another study had participants being shown three photographs: 'The Psychology of Curiosity', George Lowenstein, *Psychological Bulletin*,1994, Vol. 116. No. 1. pp. 75–98.

Curiosity is shaped like a lowercase n: 'The Psychology and Neuroscience of Curiosity', Celeste Kidd and Benjamin Y. Hayden, *Neuron*, 4 November 2015: 88(3): 449–460.

In his paper, 'The Psychology of Curiosity': 'The Psychology of Curiosity', George Lowenstein, *Psychological Bulletin*, 1994, Vol. 116. No. 1. pp. 75–98.

Mystery, he's said, 'is the catalyst for imagination': J. J. Abrams, 'The Mystery Box', TED talk, March 2007.

1.3

Consider that whole beautiful world around you, with all its: 'Exploring the Mysteries of the Brain', Gareth Cook, *Scientific American*, 6 Oct 2015.

If you hold out your arm and look at your thumbnail: *The Brain*, Michael O'Shea (Oxford University Press, 2005) p. 5.

the rest of your sight is fuzzy: *Incognito*, David Eagleman (Canongate, 2011) pp. 7–370.

blink 15 to 20 times a minute: 'Why Do We Blink so Frequently?', Joseph Stromberg, *Smithsonian*, 24 Dec 2012.

four to five saccades every second: Susan Blackmore, *Consciousness* (Oxford University Press, 2005) p. 57.

Modern filmmakers mimic saccadic behaviour: T. J. Smith, D. Levin & J. E. Cutting, 'A window on reality: Perceiving edited moving images', *Current Directions in Psychological Science*, 2012, Vol. 21, pp. 107–113.

Half didn't spot a man in a gorilla suit walk directly into the middle of the screen: Daniel J. Simons, Christopher F. Chabris, Gorillas in our midst: sustained inattentional blindness for dynamic events, *Perception*, 1999, Vol. 28, pp. 1059–1074

Other tests have confirmed we can also be: 'Beyond the Invisible Gorilla', Emma Young, *The British Psychological Research Digest*, 30 August 2018.

In a test of a simulated vehicle stop: Daniel J. Simons and Michael D. Schlosser, 'Inattentional blindness for a gun during a simulated police vehicle stop', *Cognitive Research: Principles and Implications*, 2017, 2:37.

Dr Todd Feinberg writes of a patient, Lizzy: *Altered Egos: How the Brain Creates the Self* (Oxford University Press, 2001) pp. 28–9.

less than one ten trillionth of light spectrum: *Incognito*, David Eagleman (Canongate, 2011) p. 100.

Evolution shaped us with perceptions that allow us to survive' . . . Professor Donald Hoffman has said: *The Case Against Reality*, Amanda Gefter, *The Atlantic*, 25 April 2016.

mantis shrimp: *Deviate*, Beau Lotto (Hachette 2017). Kindle location 531.

bees' eyes are able to see: *Deviate*, Beau Lotto (Hachette 2017). Kindle location 538.

Russians are raised: *How Emotions Are Made*, Lisa Feldman-Barrett (Picador 2017) p. 146.

in order to identify ripe fruit: 'You can thank your fruit-hunting ancestors for your color vision', Michael Price, *Science*, 19 Feb 2017.

Dreams feel real: *Head Trip*, Jeff Warren (Oneworld, 2009) p. 38.

to explain a 'myoclonic jerk': *Head Trip*, Jeff Warren (Oneworld, 2009) p. 31.

Wherever studies have been done: *The Storytelling Animal*, Jonathan Gottschall (HMH, 2012) p. 82.

seems to have caught people in the act of 'watching' the models of stories: *Louder than Words*, Benjamin K. Bergen (Basic, 2012) p. 63. Surprisingly, related studies suggest the brain doesn't make much distinction between stories told in the first ('I') and third singular persons ('he' or 'she'). Given sufficient context, it tends to take the 'observer perspective', as if it's watching the action of the story remotely.

It 'appears to modulate what part of an evoked simulation someone': *Louder than Words*, Benjamin K. Bergen (Basic, 2012) p. 118.

This is perhaps why transitive construction: *Louder than Words*, Benjamin K. Bergen (Basic, 2012) p. 99.

For the same reason, active sentence construction: *Louder than Words*, Benjamin K. Bergen (Basic, 2012) p. 119.

to make vivid scenes, three specific qualities: 'Differential engagement of brain regions within a "core" network during scene construction', Jennifer Summerfield, Demis Hassabis & Eleanor Maguire, *Neuropsychologia*, 2010, Vol. 48, 1501–1509.

As C. S. Lewis implored a young writer in 1956: http://www.lettersof-note.com/2012/04/c-s-lewis-on-writing.html

Only that way: A final lesson from the model-making brain is that simplicity is also crucial. The human beam of attention is narrow. 'Everything about our hominin past,' writes the neurobiologist Professor Robert Sapolsky, 'has honed us to be responsive to one face at a time.' We have hunter-gatherer brains, specialised to focus on a single moving prey animal, a single ripe fruit or a single tribal confederate. This narrowness is why stories often begin simply, from the perspective of one person, or are centred around one problem.

1.5

it's been argued: *The Domesticated Brain*, Bruce Hood (Pelican, 2014).

their physical strength as much as halving: 'The Domestication of Human', Robert G. Bednarik, 2008, *Anthropologie* XLVI/1 pp. 1–17.

Whereas ape and monkey parents: *Evolutionary Psychology*, Robin Dunbar, Louise Barrett, & John Lycett (Oneworld, 2007) p. 62.

Newborns are attracted to human faces more than to any other object: *On the Origin of Stories*, Brian Boyd (Harvard University Press, 2010) p. 96.

One hour from birth, begin imitating them: *On the Origin of Stories*, Brian Boyd (Harvard University Press, 2010) p. 96.

By two, they've learned to control their social worlds by smiling: *The Self Illusion*, Bruce Hood (Constable and Robinson, 2011) p. 29.

so adept at reading people that they're making calculations about status and character automatically, in one tenth of a second: 'Effortless Thinking', Kate Douglas, *New Scientist*, 13 December 2017.

'Our species has conquered the Earth because': *Mindwise*, Nicholas Epley (Penguin, 2014) p. xvii.

Studies indicate that those who anthropomorphise: *Mindwise*, Nicholas Epley (Penguin, 2014) p. 65.

Bankers project human moods: *Mindwise*, Nicholas Epley (Penguin. 2014) p. 62. It says much about the brain's natural storytelling instincts that these processes seem especially active when things go wrong. Whether it's a car or a computer, the more it fails, the more likely its owners are to treat is as if it has 'a mind of its own'. Epley had such owners undergo brain scans. 'We found the same neural regions involved in thinking about the minds of other people were also engaged when thinking about these unpredictable gadgets,' he writes. When trouble strikes, when the brain's predictions fail, we switch into story mode. Our narrow band of attention turns on. We become aware. And there we are, one mind primed for action in the fairytale realm of others.

Charles Dickens, William Blake and Joseph Conrad all spoke of: 'Introduction of Writer's Inner Voices', Charles Fernyhough, 4 June 2014, http://writersinnervoices.com.

the novelist and psychologist Professor Charles Fernyhough: 'Fictional characters make "experiential crossings" into real life, study finds', Richard Lea, *Guardian*, 14 Feb 2017.

some research suggests strangers read another's thoughts: *Mindwise*, Nicholas Epley (Penguin, 2014) p. 9.

Alexander Mackendrick writes, 'I start by asking': *On Film-Making*, Alexander Mackendrick (Faber & Faber, 2004) p. 168.

1.6

recent research suggests we're more likely to attend to: 'Meaning-based guidance of attention in scenes as revealed by meaning maps', John M. Henderson & Taylor R. Hayes, *Nature, Human Behaviour*, 2017, Vol. 1, pp. 743–747.

1.7

when we drink: *Subliminal*, Leonard Mlodinow (Penguin, 2012) p. 24.

The way food is described: *Subliminal*, Leonard Mlodinow (Penguin, 2012) p. 21.

use around one metaphor for every ten seconds of speech: *I Is an Other*, James Geary (Harper Perennial, 2012) p. 5.

Neuroscientists are building a powerful case: *Louder than Words*, Benjamin K. Bergen (Basic, 2012) pp. 196–206.

When participants in one study read the words, 'he had a rough day':

'Metaphorically feeling: Comprehending textural metaphors activates somatosensory cortex', Simon Lacey, Randall Stilla, K. Sathian, *Brain and Language*, Vol. 120, Issue 3, March 2012, pp. 416–421.

In another, those who read 'she shouldered the burden': 'Engagement of the left extrastriate body area during body-part metaphor comprehension', Simon Lacey, Randall Stilla, Gopikrishna Deshpande, Sinan Zhao, Careese Stephens, Kelly McCormick, David Kemmerer, K. Sathian, *Brain & Language*, 2017, 166, 1–18.

It won't come as much of a surprise to discover: *Politics and the English Language*, George Orwell (Penguin, 1946).

Researchers recently tested this idea that clichéd metaphors: *Louder than Words*, Benjamin K. Bergen (Basic, 2012) p. 206.

1.8

In a classic 1932 experiment, the psychologist Frederic Bartlett: *Subliminal*, Leonard Mlodinow (Penguin, 2012), p. 68.

Estimates vary, but it's believed the brain processes around 11 million bits: *Strangers to Ourselves*, Timothy D. Wilson, (Belknap Harvard, 2002), p.24.

no more than forty: *The Social Animal*, David Brooks (Short Books, 2011) p. x.

the 'Cosmic Hunt' myth: 'The Evolution of Myths', Julien d'Huy, *Scientific American*, December 2016.

BANANAS. VOMIT: *Thinking, Fast and Slow*, Daniel Kahneman (Penguin, 2011) p. 50.

the early twentieth century by the Soviet filmmakers: *Film Technique and Film Acting*, Vsevolod Pudovkin (Grove Press, 1954) p. 140. According to some accounts, the third shot was actually an attractive woman reclining on a chaise longue, with the audience projecting lust into the actor. In the 1954 translation of his book *Film Technique and Film Acting*, Pudovkin describes the bear.

You want all your scenes to have a 'because': Accessed at: https://johnaugust.com/2012/scriptnotes-ep-60-the-black-list-and-a-stack-of-scenes-transcript

Full quote: 'You want all your scenes to have a "Because" between them and not an "And Then" between them. And it's something that you learn and get better at which is having everything cause everything, and everything build on everything. But I have noticed, particularly in the action genre, it seems like things have gotten very episodic.'

strongly predicts an interest in poetry and the arts: *Personality*, Daniel Nettle (Oxford University Press, 2009) p. 190.

CHAPTER TWO

2.0

Mr B . . . writes the neuroscientist Professor Michael Gazzaniga: *The Consciousness Instinct*, Michael Gazzaniga (Farrahr, Straus and Giroux, 2018) pp. 136–138.

The brain constructs its hallucinated model: *Six Impossible Things Before Breakfast*, Lewis Wolpert (Faber & Faber, 2011) pp. 36–38.

The mythologist Joseph Campbell said: *The Power of Myth*, Joseph Campbell with Bill Moyers (Broadway Books, 1998) p. 3.

2.1

this personality is likely to remain relatively stable: 'A Coordinated Analysis of Big-Five Trait Change Across 16 Longitudinal Samples', Elieen Graham et al. PrePrint: https://psyarxiv.com/ryjpc/.

fictional characters. One academic paper: 'The Five-Factor Model in Fact and Fiction', Robert R. McCrae, James F. Gaines, Marie A. Wellington, 2012, 10.1002/9781118133880.hop205004.

Different personalities have different go-to tactics: In *Personality Psychology*, Larsen, Buss & Wisjeimer (McGraw Hill, 2013), a 'taxonomy of eleven tactics of manipulation' has been compiled (p. 427).

Charm ('I try to be loving when I ask her to do it')

Coercion ('I yell at him until he does it')

Silent treatment ('I don't respond to her until she does it')

Reason ('I will explain why I want him to do it')

Regression ('I whine until she does it')

Self-abasement ('I act submissive so that he will do it')

Responsibility invocation ('I get her to make a commitment to doing it')

Hardball ('I hit him so that he will do it')

Pleasure induction ('I show her how much fun it will be to do it')

Social comparison ('I tell him that everyone else is doing it')

Monetary reward ('I offer her money so that she will do it')

writes the psychologist Professor Keith Oatley: *Such Stuff as Dreams*, Keith Oatley (Wiley-Blackwell, 2011) p. 95.

Conscientious people tend to: *Personality Psychology*, Larsen, Buss & Wisjeimer (McGraw Hill, 2013) p. 69.

extraverts are more likely to have affairs: 'Sextraversion', Dr David P. Schmidt, *Psychology Today*, 28 June 2011.

and car accidents: *Personality Psychology*, Larsen, Buss & Wisjeimer (McGraw Hill, 2013) p. 68.

disagreeable people are better at fighting: *Personality*, Daniel Nettle (Oxford University Press, 2009) p. 177.

those high in openness are more likely to get tattoos: *Personality Psychology*, Larsen, Buss & Wisjeimer (McGraw Hill, 2013) p. 70.

be unhealthy: *Snoop*, Sam Gosling (Basic Books, 2008) p. 99

and vote for left-wing political parties: *Personality Psychology*, Larsen, Buss & Wisjeimer (McGraw Hill, 2013) p. 70.

while those low in conscientiousness are more likely to end up in prison: *Personality Psychology*, Larsen, Buss & Wisjeimer (McGraw Hill, 2013) p. 69.

and have a higher risk of dying: *Personality*, Daniel Nettle (Oxford University Press, 2009) p. 34.

males tend to be more disagreeable than females: *Personality*, Daniel Nettle (Oxford University Press, 2009) p. 177. Nettle quotes the 70 per cent figure but I was warned by one of my expert proof readers, Dr Stuart Ritchie, that although the study Nettle quotes is robust, other robust studies find less dramatic scores. 60 per cent was agreed to be a safer figure to quote.

A similar personality gap is found for neuroticism: Comments from Dr Stuart Ritchie.

2.2

'Zaha Hadid', Lynn Barber, *Observer*, 9 March 2008.

'Human personalities are rather like fractals,': *Personality*, Daniel Nettle (Oxford University Press, 2009) p. 7.

People make 'identity claims': *Snoop*, Sam Gosling (Basic Books, 2008) pp. 12–19.

The psychologist Professor Sam Gosling advises: *Snoop*, Sam Gosling (Basic Books, 2008) p. 19.

2.4

Between the ages of zero and two: *The Social Animal*, David Brooks (Short Books, 2011) p. 47.

It's the main reason we have such greatly extended childhoods: *The Self Illusion*, Bruce Hood (Constable and Robinson, 2011) p. 22.

Play, including storytelling, is typically overseen: *Brain and Culture*, Bruce Wexler (MIT Press, 2008) p. 134. See also: C. M. Walker & T. Lombrozo, 'Explaining the moral of the story', *Cognition*, 2017, 167, 266–281.

One study into the backgrounds of sociopathic: 'A History of Children's Play and Play Environments', Joe L. Frost (Routledge, 2009) p. 208.

It's in our first seven years: 'The Construction of the Self', Susan Harter (Guildford Press, 2012) p. 50.

According to some psychologists: 'The Geography of Thought', Richard E. Nisbett (Nicholas Brealey, 2003). A fuller exploration of these ideas features in my book *Selfie* (Picador, 2017) in Book Two: The Perfectible Self.

Because individual self-reliance was the key to success: These differences remain widespread today. If you show an Asian student a cartoon of a fish tank and track their saccades by the millisecond, they unconsciously scan the entire scene, while their Western counterpart focuses more on the dominant, individual, brightly coloured fish at the front. Ask what they saw and the Asian description is more likely to begin with the context – 'I saw a tank' – compared to the Westerner's individual object – 'I saw a fish'. Ask what they thought of that singular fish and the Westerner is likely to say 'it was the leader' whilst the Easterner assumes it's done something wrong because it's excluded from the group.

Such cultural differences create radically different experiences of life, self and story. When asked to draw a 'sociogram' of themselves in relation to everyone they know, Westerners tend to draw themselves as a big circle in the middle, while Easterners tend to make themselves small, towards the edge. In China, unlike the West, humble and hardworking students are popular, whilst shyness is considered a leadership quality. Such differences begin in the neural models and therefore control our perception of reality. 'It isn't just that Easterners versus Westerners think about the world differently,' the psychologist Professor Richard Nisbett told me. 'They're literally seeing a different world.' This can trigger serious conflicts, with one side simply not perceiving moral realities that seem obvious to the other. 'The Chinese are willing to accept the idea of unjustly punishing someone if that makes the group better off,' Nisbett said. 'That's an outrage to Westerners who are so individual-rights orientated. But, to them, the group is everything.'

It 'changed the way people thought about cause and effect': 'Life on Purpose', Victor Stretcher (Harper One, 2016) p. 24.

one three-year-old girl in the US: *The Storytelling Animal*, Jonathan Gottschall (HMH, 2012) p. 33.

practically no real autobiography for two thousand years: *The Autobiographical Self in Time and Culture*, Qi Wang (Oxford University Press, 2013) pp. 46, 52.

according to the psychologist Professor Uichol Kim: Interview with author.

2.5

we endeavour to understand our life as a 'grand narrative': *The Redemptive Self*, Dan P. McAdams (Oxford University Press, 2013) p. xii the neurobiologist Professor Bruce Wexler describes it: *Brain and Culture*, Bruce Wexler (MIT Press, 2008) p. 9.

As Wexler writes: *Brain and Culture*, Bruce Wexler (MIT Press, 2008) p. 9.

the 'makes sense stopping rule': *The Happiness Hypothesis*, Jonathan Haidt (Arrow, 2006) p. 65.

Not only do our neural reward systems spike pleasurably: *The Political Brain*, Drew Westen (Public Affairs, 2007) pp. x–xiv.

It's not simply that we ignore or forget evidence: A fuller exploration into confirmation bias features in my book *The Heretics* (Picador, 2013), in chapter six: 'The Invisible Actor at the Centre of the World'.

Smart people are mostly better: 'Myside Bias, Rational Thinking, and Intelligence', Keith E. Stanovich, Richard F. West, Maggie E. Toplak, *Current Directions in Psychological Science*, 2013, Vol. 22, Issue 4.

'Cognitive Sophistication Does Not Attenuate the Bias Blind Spot', Richard F. West, Russell J. Meserve, and Keith E. Stanovich, *Journal of Personality and Social Psychology*, 4 June 2012.

One compelling theory: This is the thesis of *The Enigma of Reason* by Hugo Marcier and Dan Sperber (Allen Lane, 2017).

the screenwriter Russell T. Davies's observation: 'Has every conversation in history been just a series of meaningless beeps?', Charlie Brooker, *Guardian*, 28 April 2013.

Things are experienced as pleasurable: *Brain and Culture*, Bruce Wexler (MIT Press, 2008) p. 9.

The neuroscientist Sarah Gimbel watched what happened: 'You Are Not So Smart with David McRaney', *The Neuroscience of Changing Your Mind*, Episode 93, 13 Jan 2017.

2.6

Among the most powerful of these beliefs: 'The Illusion of Moral Superiority', B. M. Tappin, R. T. McKay, *Soc Psychol Personal Sci*, 2017, Aug 8(6):623–631.

participants split money with anonymous others: 'Motivated misremembering: Selfish decisions are more generous in hindsight',

Ryan Carlson, Michel Marechal, Bastiaan Oud, Ernst Fehr, Molly Crockett, 23 July 2018. PrePrint accessed at: https://psyarxiv.com/7ck25/

What is selected as a personal memory: 'The "real you" is a myth – we constantly create false memories to achieve the identity we want', Giuliana Mazzoni, *The Conversation*, 19 Sept 2018.

Work by Mazzoni and others: 'Changing beliefs and memories through dream interpretation', Giuliana A. L. Mazzoni, Elizabeth F. Loftus, Aaron Seitz, Steven J. Lynn, *Applied Cognitive Psychology*, Vol. 13, Issue 2, April 1999, pp. 125–144.

For the psychologists Professors Carol Tavris and Elliot Aronson: *Mistakes Were Made (But Not By Me)*, Carol Tavris and Elliot Aronson (Pinter and Martin, 2007) p. 76.

Professor Nicholas Epley catches this hero-maker lie: *Mindwise*, Nicholas Epley (Penguin, 2014) p. 54.

Moral superiority is thought to be: 'The Illusion of Moral Superiority', B. M. Tappin, R. T. McKay, *Soc. Psychol Personal Sci*, 2017, Aug;8(6): 623–631.

Maintaining a 'positive moral self-image': 'Motivated misremembering: Selfish decisions are more generous in hindsight', Ryan Carlson, Michel Marechal, Bastiaan Oud, Ernst Fehr, Molly Crockett, 23 July 2018. PrePrint accessed at: https://psyarxiv.com/7ck25.

Even murderers and domestic abusers: *The Happiness Hypothesis*, Jonathan Haidt (Heinemann, 2006) p. 73.

When researchers tested prisoners: 'Behind bars but above the bar: Prisoners consider themselves more prosocial than non-prisoner', Constantine Sedikides, Rosie Meek, Mark D. Alicke and Sarah Taylor, *British Journal of Social Psychology*, 2014, 53, 396–403.

as did Hitler, whose last words: *Hitler's World View: A Blueprint for Power*, Eberhard Jäckel (Harvard University Press, 1981) p. 65.

One 35-year-old metal worker, remembered: *Ordinary Men*, Christopher R. Browning (Harper Perennial, 2017) p. 73.

Researchers have found that violence and cruelty: *The Happiness Hypothesis*, Jonathan Haidt (Heinemann, 2006) p. 75.

2.7

One such real-life hero is the former 'eco-terrorist' Mark Lynas: Interview with author.

CHAPTER THREE

3.0

Lisa Bortolotti explains: 'Confabulation: why telling ourselves stories makes us feel OK', Lisa Bortolotti, *Aeon*, 13 February 2018.

series of famous experiments: My account of Gazzaniga's confabulation experiments is sourced from his books *Who's In Charge?* (Robinson, 2011) and *Human* (Harper Perennial, 2008). Another excellent telling can be found in *The Happiness Hypothesis*, Jonathan Haidt (Heinemann, 2006).

The job of the narrator, writes Gazzaniga: *Who's in Charge?*, Michael Gazzaniga (Robinson, 2011) p. 85.

It's because of such findings: *Mindwise*, Nicholas Epley (Penguin, 2014) p. 30.

Leonard Mlodinow said years of psychotherapy: *Subliminal*, Leonard Mlodinow, (Penguin, 2012) p. 177.

3.1

we're a riotous democracy of mini-selves: *Incognito: The Secret Lives of the Brain*, David Eagleman (Canongate, 2011) p. 104.

Fabrication of stories, he adds: *Incognito: The Secret Lives of the Brain*, David Eagleman (Canongate, 2011) p. 137.

Kurt Goldstein recalled a woman whose left hand: *Altered Egos: How the Brain Creates the Self*, Todd E. Feinberg (Oxford University Press, 2001) pp. 93–99.

Todd Feinberg saw a patient whose hand: *Altered Egos: How the Brain Creates the Self*, Todd E. Feinberg (Oxford University Press, 2001) pp. 93–99.

The BBC told of a patient: 'Alien Hand Syndrome sees woman attacked by her own hand', Dr Michael Mosley, 20 January 2011.

grabbed his wife with his left hand: *Altered Egos: How the Brain Creates the Self*, Todd E. Feinberg (Oxford University Press, 2001) pp. 93–99.

A child can't consciously accept: *The Uses of Enchantment*, Bruno Bettelheim (Penguin, 1976) p. 30.

all the child's wishful thinking: *The Uses of Enchantment*, Bruno Bettelheim (Penguin, 1976) p. 66.

They operate 'in two realms': *Making Stories*, Jerome Bruner (Harvard University Press, 2002) p. 26.

the psychologist Professor Brian Little writes: *Who Are You Really?*, Brian Little (Simon & Schuster, 2017) p. 25.

3.4

Robert McKee writes: *Story*, Robert McKee (Methuen, 1999) p. 138.

3.6

We've spent more than ninety-five per cent: *Who's In Charge?*, Michael Gazzaniga (Robinson, 2011) p. 315.

we still have Stone Age brains: *Grooming, Gossip and the Evolution of Language*, Robin Dunbar (Faber & Faber, 1996), Kindle Locations 1255–1256.

people prefer to sleep as far from their bedroom door: *Evolutionary Psychology*, David M. Buss (Routledge, 2016) p. 84.

The body's reflexes remain primed: *The Origins of Creativity*, Edward O. Wilson (Liveright, 2017) p. 114.

All over the world, people enjoy open spaces: *Evolutionary Psychology*, David M. Buss (Routledge, 2016) p. 84.

psychologists argue that human language: *Evolutionary Psychology*, Robin Dunbar, Louise Barrett, John Lycett (Oneworld, 2007) p. 133.

Human tribes were big: *Grooming, Gossip and the Evolution of Language*, Robin Dunbar (Faber & Faber, 1996), Kindle Locations 1152–1156.

occupy a large physical territory: *Evolutionary Psychology* by Robin Dunbar, Louise Barrett, John Lycett (Oneworld, 2007) p. 112.

'Stories arose out of our intense interest in social monitoring': *On The Origin of Stories*, Brian Boyd (Harvard University Press, 2010) p. 64.

An analysis of ethnographic accounts: O. S. Curry, D. A. Mullins, H. Whitehouse. Is it good to cooperate? 'Testing the theory of morality-as-cooperation in 60 societies', *Current Anthropology*, 15 July 2017.

Even pre-verbal babies: *Just Babies*, Paul Bloom (Bodley Head, 2013) p. 27.

Psychologist Professor Paul Bloom writes: *Just Babies*, Paul Bloom (Bodley Head, 2013) p. 27.

Joseph Campbell describes: *The Power of Myth*, Joseph Campbell with Bill Moyers (Broadway Books, 1998) p. 126.

Christopher Booker writes that: *The Seven Basic Plots*, Christopher Booker (Continuum, 2005) p. 555.

Another psychologist's puppet show: *The Domesticated Brain*, Bruce Hood (Pelican, 2014) p. 195.

Brain scans reveal: *Comeuppance*, William Flesch (Harvard University Press, 2009) p. 43.

a form of what's known as 'costly signalling': *Grooming, Gossip and the Evolution of Language*, Robin Dunbar (Faber & Faber, 1996), Kindle Locations 2911–2917.

'The heroes and heroines of narrative': *Comeuppance*, William Flesch (Harvard University Press, 2009) p. 126.

not only is gossip universal: *Moral Tribes*, Joshua Greene (Atlantic Books, 2013) p. 45. Gossip-type behaviour has even been shown in three-year-olds: Preschoolers affect others' reputations through prosocial gossip: http://onlinelibrary.wiley.com/doi/10.1111/bjdp.12143/abstract?campaign=woletoc.

most of it concerns moral infractions: *Just Babies*, Paul Bloom (Bodley Head, 2013) p. 95.

3.7

Evolutionary psychologists argue: *The Redemptive Self*, Dan P. McAdams (Oxford University Press, 2013) p. 29.

Getting ahead means gaining status: 'Is the Desire for Status a Fundamental Human Motive? A Review of the Empirical Literature', C. Anderson, J. A. D. Hildreth & L. Howland, *Psychological Bulletin*, 16 March 2015.

'Humans naturally pursue status': *On the Origin of Stories*, Brian Boyd (Harvard University Press, 2010) p. 109.

people's 'subjective well-being, self-esteem': 'Is the Desire for Status a Fundamental Human Motive? A Review of the Empirical Literature', C. Anderson, J. A. D. Hildreth & L. Howland, *Psychological Bulletin*, 16 March 2015.

Studies of gossip in contemporary hunter-gatherer tribes: *Behave*, Robert Sapolsky (Vintage, 2017) p. 323.

Even crickets keep a tally: *Evolutionary Psychology*, David M. Buss (Routledge, 2016) p. 49.

the astonishing fact that not only do ravens: *Behave*, Robert Sapolsky (Vintage, 2017) p. 428.

have a lifespan at the top of about four to five years: *Our Inner Ape*, Frans de Waal (Granta, 2005) p. 68.

benefits for chimps and humans include: *Comeuppance*, William Flesch (Harvard University Press, 2009) p. 110.

'The tendency of chimps to rally for the underdog': *Our Inner Ape*, Frans de Waal (Granta, 2005) p. 75. Of course, humans, too, root for the underdog: *The Appeal of the Underdog*, Joseph A. Vandello, Nadav P. Goldschmied and David A. R. Richards, *Pers Soc Psychol Bull*, 200, 33: 1603.

Christopher Booker writes: *The Seven Basic Plots*, Christopher Booker (Continuum, 2005) p. 556.

a 'hero and heroine must represent: *The Seven Basic Plots*, Christopher Booker (Continuum, 2005) p. 268.

Biographer Tom Bower writes: 'The pampered, petulant, self-pitying Prince', Tom Bower, *Daily Mail*, 16 March 2018.

When people in brain scanners: *Behave*, Robert Sapolsky (Vintage. 2017) p. 67.

When they read about them suffering a misfortune: *Behave*, Robert Sapolsky (Vintage, 2017) p. 67.

researchers at Shenzhen University: 'Social hierarchy modulates neural responses of empathy for pain', Chunliang Feng, Zhihao Li, Xue Feng, Lili Wang, Tengxiang Tian, Yue-Jia Luo, *Social Cognitive and Affective Neuroscience*, Vol. 11, Issue 3, 1 March 2016, pp. 485–495.

A study of over 200 popular nineteenth- and early twentieth-century novels: *Palaeolithic Politics in British Novels of the Longer Nineteenth Century*, Joseph Cattoll et al., accessed at: http://www.personal.psu.edu/~j5j/papers/PaleoCondensed.pdf

3.8

But on this 'raw and gusty' day, Caesar failed: *Such Stuff as Dreams*, Keith Oatley (Wiley-Blackwell, 2011) p. 94.

Psychologists define humiliation: 'Humiliation: its Nature and Consequences', Walter J. Torres and Raymond M. Bergner, *Journal of the American Academy of Psychiatry and the Law Online*, June 2010, 38 (2) 195–204.

As Professor William Flesch writes: *Comeuppance*, William Flesch (Harvard University Press, 2009) p. 159.

3.9

Babylon, 587 BC. A group of 4,000 high-status men: *The Written World*, Martin Puchner (Granta 2017) pp. 46–59.

'They immediately bowed their heads to the ground': *The Written World*, Martin Puchner (Granta 2017) p. 54.

A recent study of eighteen hunter-gatherer tribes: 'Cooperation and the evolution of hunter-gatherer storytelling', Daniel Smith et al., *Nature Communications*, Volume 8, Article number: 1853, 5 December 2017,

'We all belong to multiple in-groups': *Subliminal*, Leonard Mlodinow (Penguin, 2012) p. 165.

Tribal stories blind us: *The Political Brain*, Drew Westen (Public Affairs, 2007) p. xvi.

Jonathan Haidt has explored: Capitalism is Exploitation: https://www.youtube.com/watch?v=9B-RkNRGH9s

Capitalism is Liberation: https://www.youtube.com/watch?v=kOomUpEdLE4&list=UUFHCypPBiy5cpLKFX11q0QQ

In the twentieth century alone: *Our Inner Ape*, Frans de Waal (Granta, 2005) p. 5.

halting in silence: *Our Inner Ape*, Frans de Waal (Granta, 2005) p.132.

When caught, a 'foreign' chimp is savagely beaten to death: *Our Inner Ape*, Frans de Waal (Granta, 2005) pp. 24, 132.

'it cannot be coincidental that the only animals': *Our Inner Ape*, Frans de Waal (Granta, 2005) p. 137.

It takes its individuals and erases their depth and diversity: 'Intergroup Perception in the Social Context: The Effects of Social Status and Group Membership on Perceived Out-Group Homogeneity', Markus Brauer, *Journal of Experimental Social Psychology*, 37 (2001): 15–31.

caused viewers to pour en masse into the streets of Berlin: 'Jud Süss: The Film That Fuelled the Holocaust', Gary Kidney, *Warfare History Network*, 23 March 2016.

from spree shootings to honour killings: *The Domesticated Brain*, Bruce Hood (Pelican, 2014) p. 278; *Behave*, Robert Sapolsky (Vintage 2017) p. 288.

Many deploy a third incendiary group emotion: disgust: 'Evil Origins: A Darwinian Genealogy of the Popcultural Villain', J. Kjeldgaard-Christiansen, *Evolutionary Behavioral Sciences*, 2015, 10(2), 109–122.

3.10

In his inquiry into the psychology of fairy tales: *The Uses of Enchantment*, Bruno Bettelheim (Penguin, 1976) p. 10.

3.11

infants whose caregivers behave unpredictably: *The Domesticated Brain*, Bruce Hood (Pelican, 2014) p. 116.

The body has a dedicated network of touch receptors: 'Why your brain needs touch to make you human', Linda Geddes, *New Scientist*, 25 February 2015.

doesn't merely alter who we are as adults superficially: *The Popularity Illusion*, Mitch Prinstein (Penguin, 2018) Kindle location 1984.

those who'd had high-school experiences of loneliness: *The Popularity Illusion*, Mitch Prinstein (Penguin, 2018) Kindle location 2105.

Similar tests had people viewing simple cartoons: *The Popularity Illusion*, Mitch Prinstein (Penguin, 2018) Kindle location 2111.

CHAPTER FOUR

4.0

For the nineteenth-century critic Ferdinand Brunetière: *On Film-Making*, Alexander Mackendrick (Faber & Faber, 2004) p. 106

'almost as basic a need as': *The Happiness Hypothesis*, Jonathan Haidt (Arrow, 2006) p. 22.

When researchers put people in flotation tanks: *Brain and Culture*, Bruce Wexler (MIT Press, 2008) pp. 76–77.

Another study found 67 per cent of male participants: 'Just Think: The challenges of the Disengaged Mind', Timothy D. Wilson et al., *Science*, July 2014, 345(6192), pp. 75–7.

John Bransford and Marcia Johnson: *The Sense of Style*, Steven Pinker (Penguin, 2014) p. 147.

One clever study asked restaurant employees to circle: *Mindwise*, Nicholas Epley (Penguin, 2014) p. 50.

Another test found that eight in every ten: *The Domesticated Brain*, Bruce Hood (Pelican, 2014) p. 222.

using a language millions of years older: *The Political Brain*, Drew Westen (Public Affairs, 2007) p. 57.

Daniel Nettle writes: *Personality*, Daniel Nettle (Oxford University Press, 2009) p. 87.

One Welsh teenager, Jamie Callis: 'The real-life story of a computer game addict who played for up to 16 hours a day by Mark Smith', *Wales Online*, 18 Sept 2018.

In South Korea, two parents: 'S Korea child starves as parents raise virtual baby', BBC News, 5 March 2010.

a mixture of 'trivial pursuits and magnificent obsessions': *Who Are You Really?*, Brian Little (Simon & Schuster, 2017) p. 45.

Aristotle contemptuously dismissed the hedonists: *Life on Purpose*, Victor Stretcher (Harper One, 2016) p. 27.

'It's living in a way that fulfils our purpose': Interview with author.

living with a sufficient sense of purpose reduces: 'A meaning to life: How a sense of purpose can keep you healthy', Teal Burrell, *New Scientist*, 25 Jan 2017.

Results from a team led by Professor of medicine Steve Cole: I wrote about Steve Cole's work in the *New Yorker* ('A Better Kind of Happiness', 7 July 2016).

'Some people wander aimlessly through life': 'Purpose in Life as a Predictor of Mortality Across Adulthood', Patrick Hill and Nicholas Turiano, *Psychological Science*, May 2014, 25(7) pp. 1487–96.

Our reward systems spike: Video lecture: 'Dopamine Jackpot! Sapolsky on the Science of Pleasure', http:// www.dailymotion.com/video/xh6ceu_ dopamine-jackpot-sapolsky- on-the-science-of-pleasure_news.

the words 'do', 'need' and 'want': *The Bestseller Code*, Jodie Archer & Matthew L. Jockers (Allen Lane, 2016) p. 163.

4.1

Researchers downloaded 1,327: 'The emotional arcs of stories are dominated by six basic shapes', Andrew J. Reagan, Lewis Mitchell, Dilan Kiley, Christopher M. Danforth, Peter Sheridan Dodds, *EPJ Data Science*, 5:31, 4 November 2016.

4.2

For the neuroscientist Professor Beau Lotto: *Deviate*, Beau Lotto (W&N, 2017) Kindle location 685.

When the data scientist David Robinson: Examining the arc of 100,000 stories: a tidy analysis by David Robinson, http://varianceexplained. org/r/tidytext-plots, 26 April 2017.

The psychologist and story analyst Professor Jordan Peterson: Maps of Meaning video lectures. Jordan Peterson, 2017: Marionettes & Individuals Part Three [01:35]

There's the second level in which characters are altered: Story analysts disagree on the nature of character change. Some say protagonists *transform* their essential character, others that they *reveal* some part that was previously hidden. Both positions have merit. When characters change, they're forcing a better subconscious model of self into dominance, reinforcing the neural networks that conjure this self into being, so it more often wins the neural debates that ultimately control the character's behaviour.

In doing so the characters *expand* who they are, giving themselves greater elasticity around their core personality, which gives them a more varied collection of tools for controlling the world of humans.

For simplicity's sake, our focus has been on the changeful journey of an individual protagonist. But, it hopefully hardly needs to be said that *all* the significant characters in story go through journeys of change, albeit possibly in ways subordinate to a protagonist. They're all asked that subconscious question until the plot is done with them. They all keep changing. Those changes probably won't be linear. They'll move back and forth and up and down. But the change never stops. An immersive plot is a complex and beautiful symphony of change, because brains are obsessed by change.

4.3

When study participants were faced with a machine: *The Self Illusion*, Bruce Hood (Constable, 2011) p. 51.

Another test found that participants given electric shocks: *The Domesticated Brain*, Bruce Hood (Pelican, 2014) p. 115.

'A critical element to our well-being': *Redirect*, Timothy D. Wilson (Penguin, 2013) p. 268.

Roy Baumeister writes that: *The Cultural Animal*, Roy Baumeister (Oxford University Press, 2005) p. 102.

4.4

'the invisible actor': *Making up the Mind*, Chris Frith (Blackwell Publishing, 2007) p. 109.

'the transported "traveller" can return changed': 'The Extended Transportation-Imagery Model: A Meta-Analysis of the Antecedents and Consequences of Consumers' Narrative Transportation', Tom van Laer, Ko de Ruyter, Luca M. Visconti and Martin Wetzels; *Journal of Consumer Research*, Vol. 40, No. 5 (February 2014) pp. 797–817.

4.5

One study had a group of white Americans: 'Entertainment-education effectively reduces prejudice', Sohad Murrar, Markus Brauer; Group Processes & Intergroup Relation, 2018, Vol 21, Issue 7.

INDEX

GRIFFITH COLLEGE DUBLIN
SOUTH CIRCULAR ROAD DUBLIN 8.
Tel: 01 4150490 Fax: (01) 4549265
library@griffith.ie

GRIFFITH COLLEGE DUBLIN
SOUTH CIRCULAR ROAD DUBLIN 8
Tel: 01 4154600 Fax: (01) 4549265
library@gcd.ie

Y049219